COMPUTER SCIENCE, TECHNOLOGY AND APPLICATIONS

ARTIFICIAL NEURAL NETWORK MODELING OF WATER AND WASTEWATER TREATMENT PROCESSES

COMPUTER SCIENCE, TECHNOLOGY AND APPLICATIONS

Additional books in this series can be found on Nova's website under the Series tab.

COMPUTER SCIENCE, TECHNOLOGY AND APPLICATIONS

ARTIFICIAL NEURAL NETWORK MODELING OF WATER AND WASTEWATER TREATMENT PROCESSES

ALI R. KHATAEE
AND
MASOUD B. KASIRI

Nova Science Publishers, Inc.
New York

Copyright © 2011 by Nova Science Publishers, Inc.

All rights reserved. No part of this book may be reproduced, stored in a retrieval system or transmitted in any form or by any means: electronic, electrostatic, magnetic, tape, mechanical photocopying, recording or otherwise without the written permission of the Publisher.

For permission to use material from this book please contact us:
Telephone 631-231-7269; Fax 631-231-8175
Web Site: http://www.novapublishers.com

NOTICE TO THE READER

The Publisher has taken reasonable care in the preparation of this book, but makes no expressed or implied warranty of any kind and assumes no responsibility for any errors or omissions. No liability is assumed for incidental or consequential damages in connection with or arising out of information contained in this book. The Publisher shall not be liable for any special, consequential, or exemplary damages resulting, in whole or in part, from the readers' use of, or reliance upon, this material. Any parts of this book based on government reports are so indicated and copyright is claimed for those parts to the extent applicable to compilations of such works.

Independent verification should be sought for any data, advice or recommendations contained in this book. In addition, no responsibility is assumed by the publisher for any injury and/or damage to persons or property arising from any methods, products, instructions, ideas or otherwise contained in this publication.

This publication is designed to provide accurate and authoritative information with regard to the subject matter covered herein. It is sold with the clear understanding that the Publisher is not engaged in rendering legal or any other professional services. If legal or any other expert assistance is required, the services of a competent person should be sought. FROM A DECLARATION OF PARTICIPANTS JOINTLY ADOPTED BY A COMMITTEE OF THE AMERICAN BAR ASSOCIATION AND A COMMITTEE OF PUBLISHERS.

Additional color graphics may be available in the e-book version of this book.

LIBRARY OF CONGRESS CATALOGING-IN-PUBLICATION DATA

Artificial neural network modeling of water and wastewater treatment
processes / editors, Ali R. Khataee, Masoud B. Kasiri.
 p. cm.
 Includes bibliographical references and index.
 ISBN 978-1-61122-781-9 (softcover)
 1. Sewage--Purification--Computer simulation. 2.
Water--Purification--Computer simulation. 3. Neural networks (Computer
science) I. Khataee, A. R. (Ali Reza), 1977- II. Kasiri, Masoud B.
 TD745.A77 2010
 628.1'620113--dc22
                    2010041334

Published by Nova Science Publishers, Inc. † New York

CONTENTS

Preface		vii
Chapter 1	Introduction	1
Chapter 2	Topology of Artificial Neural Networks	3
Chapter 3	Training, Validation and Test of a Neural Network	17
Chapter 4	Applications of Artificial Neural Network Modeling	27
Chapter 5	ANN Modeling of Adsorption Processes	31
Chapter 6	ANN Modeling of Biological Treatment Processes	37
Chapter 7	ANN Modeling of Electrochemical Treatment Processes	47
Chapter 8	ANN Modeling of Photocatalytic Processes	55
Chapter 9	ANN Modeling of Photooxidative Processes	63
Conclusions		85
Acknowledgements		87
References		89
Index		101

PREFACE

Artificial neural networks (ANNs) are computer based systems that are designed to simulate the learning process of neurons in the human brain. ANNs have been attracting great interest during the last decade as predictive models and pattern recognition. Artificial neural networks possess the ability to "learn" from a set of experimental data (e.g. processing conditions and corresponding responses) without actual knowledge of the physical and chemical laws that govern the system. Therefore, ANNs application in data treatment is high especially where systems present nonlinearities and complex behavior.

A growing world population, unrelenting urbanization, increasing scarcity of good quality water resources and rising fertilizer applications are the driving forces behind the accelerating upward trend in the use of new and more efficient methods of water and wastewater treatment. Due to the complexity of reactions in these new processes, the kinetic parameters of the various steps involved are very difficult to determine, leading to uncertainties in the design and scale-up of chemical reactors of industrial interest. Since the treatment efficiency depends on several factors, the modeling of these processes involves many problems. It is evident that these difficulties can not be solved by simple linear multivariate correlation. As a result of good modeling capabilities, artificial neural networks have been used extensively for a number of treatment processes. One of the characteristics of modeling based on ANNs is that it does not require the mathematical description of the phenomena involved in the process. This document briefly describes the application of artificial neural networks for modeling of water and wastewater treatment processes. Examples of early applications of ANNs in modeling and simulation of electrochemical treatment, photocatalytic, advanced treatment, photooxidative and adsorption processes are reviewed.

By: Ali R. Khataee
    Department of Applied Chemistry, Faculty of Chemistry, University of Tabriz, Tabriz, Iran
    and
Masoud B. Kasiri
    Faculty of Applied Arts, Tabriz Islamic Art University, Tabriz, Iran

Chapter 1

INTRODUCTION

A first wave of interest in artificial neural network (ANN) emerged after the introduction of simplified neurons by McCulloch and Pitts in 1943 [1]. These neurons were presented as models of biological neurons and as conceptual components for circuits that could perform computational tasks. ANN models were inspired by the biological sciences which study how the neuroanatomy of living animals has developed in solving problems. In the last decades great strides have been made in neural network technology. This breakthrough has led to increasing research on a wide variety of scientific applications [2]. The interest to ANNs is reflected in the number of scientists, the amounts of funding, the number of large conferences, and the number of journals associated with neural networks.

Neural networks have been trained to perform complex functions in various fields of application including pattern recognition, identification, classification, speech, vision and control systems. In this document, we have briefly described the application of artificial neural networks for modeling of different water and wastewater treatment processes.

Conventional water and wastewater treatment processes have been long established in removing many chemical and microbial contaminants of concern to public health and the environment. However, the effectiveness of these processes has become limited over the last two decades because of the following three new challenges [3, 4]:

- Increased knowledge about the consequences of water pollution and the public desire for better quality water;

- Diminishing water resources and rapid population growth and industrial development. The reuse of municipal and industrial wastewaters and the recovery of potential pollutants used in industrial processes become more critical;
- Advances in the manufacturing industry and the growing market associated with advanced treatment processes have resulted in substantial improvements to the versatility and costs of these processes at the industrial scale.

To resolve these new challenges and better use of economical resources, various advanced treatment technologies have been proposed, tested, and applied to meet both current and anticipated treatment requirements. Advanced treatment technologies have been demonstrated to remove various potentially harmful compounds that could not be effectively removed by conventional treatment processes. Among them, advanced oxidation processes have been proven to successfully remove a wide range of challenging contaminants and hold great promise in water and wastewater treatment.

This document starts with a chapter in which the topology of artificial neural networks is discussed. In chapter 3, training, validation and test of the neural networks are described. Chapter 4 follows the different applications of the neural networks. Chapters 5 to 9 discuss the early applications of ANNs in modeling and simulation of various water and wastewater treatment technologies including electrochemical and biological treatment, photocatalytic, photooxidative and adsorption processes.

Chapter 2

TOPOLOGY OF ARTIFICIAL NEURAL NETWORKS

The success in obtaining a reliable and robust network depends strongly on the choice of process variables involved as well as the available set of data and the domain used for training purposes [5]. In this chapter, we are to describe the topology of artificial neural networks including transfer functions, learning processes and training algorithms.

There are several types of artificial neural networks. Two popular ANNs are (i) multi layer feed-forward neural network trained by back propagation algorithm that is widely used, and (ii) Kohonen self-organizing mapping [6]. Each network consists of artificial neurons grouped into layers and put in relation to each other by parallel connections. The strength of these interconnections is determined by the weight associated with them. For every ANN, the first layer constitutes the input layer (independent variables) and last one forms the output layer (dependent variables). One or more neuron layers called hidden layers can be located between them.

The topology of an artificial neural network is determined by number of its layers, number of nodes in each layer and the nature of transfer functions. Optimization of ANN topology is probably the most important step in the development of a model.

A neural network is composed of neurons, usually arranged in three layers:

- The first layer, called input layer, receives data from sources that are the subject of analysis.

- The second layer is hidden layer and an intrinsic value for the neural network and has no direct contact with the outside. The activation functions of the neurons in the hidden layer are usually nonlinear; however there is no restricted rule. The number of hidden layers and the number of neurons in the hidden layers are not predefined and must be adjusted. In general, it is better to start with medium-sized hidden layers.
- The third layer is called output layer. It gives the results obtained by the network after compilation of data entered into the first layer.

The number of input and output neurons represents effectively the number of variables used in the prediction and the number of variables to be predicted, respectively. The hidden layers act like feature detectors and in theory, there can be more than one hidden layer. Universal approximation theory, however, suggests that a network with a single hidden layer with a sufficiently large number of neurons can interpret any input–output structure [5, 7-8]. The number of neurons in the hidden layer is determined by the desired accuracy in the neural predictions. Hence, it may be considered as a parameter for the neural net design. In the feed-forward neural net, all the neurons of a particular layer are connected to all the neurons of the layer next to it. The input layer of neurons acts like a distributor and the input to this layer is directly transmitted to the hidden layer. The inputs to hidden and output layers are calculated by performing a weighted summation of all the inputs received from the preceding layer. The weighted sum of the inputs are transferred to the hidden neurons, where it is transformed using an activation function. The output of hidden neurons in turn, acts as inputs to output neurons where it undergoes another transformation. Figure 1 shows a typical feed-forward neural network with a single hidden layer containing 14 neurons [9].

As it was mentioned, each artificial neuron is an elementary processor that receives a number of signals from previous layer's neurons and each of these entries has an associated weight representing the strength of connections between correspondent neurons. This, therefore, dictates two specific characteristics of each neuron:

- A "potential" or "activation" equals to the weighted sum of inputs and,
- A transfer function that produces output of the neuron according to its "activation" force.

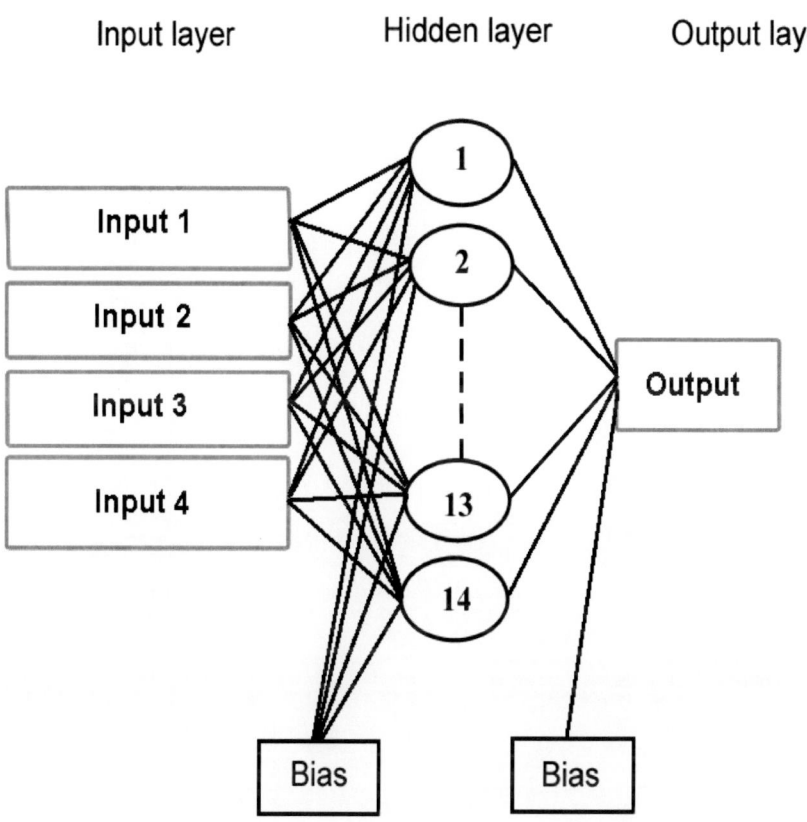

Figure 1. A typical feed-forward neural network with a single hidden layer containing 14 neurons.

2.1. TRANSFER FUNCTIONS

Different transfer functions can be used as the neuron activation function. Table 1 shows the characteristics of some of these functions. The graph in the square to the right of each transfer function represents the symbol of each transfer function.

Table 1. Characteristics of the different transfer functions

Function name	Input/Output	Icon	Acronym (*Matlab Name*)
Hard-limit	$a = 0$ if $n < 0$ $a = 1$ if $n \geq 0$		hardlim
Symmetric hard-limit	$a = -1$ if $n < 0$ $a = 1$ if $n \geq 0$		hardlims
Linear	$a = n$		purelim
Saturated linear	$a = 0$ if $n < 0$ $a = n$ if $0 \leq n \leq 1$ $a = 1$ if $n > 1$		satlin
Symmetric saturated linear	$a = -1$ if $n < -1$ $a = n$ if $-1 \leq n \leq 1$ $a = 1$ if $n > 1$		satlins
Positive linear	$a = 0$ if $n < 0$ $a = n$ if $n \geq 0$		poslin
Sigmoid	$a = \dfrac{1}{1+\exp^{-n}}$		logsig
Tangent hyperbolic	$a = \dfrac{e^n - e^{-n}}{e^n + e^{-n}}$		tansig
Competitive	$a = 1$ if n is maximum $a = 0$ otherwise		compet

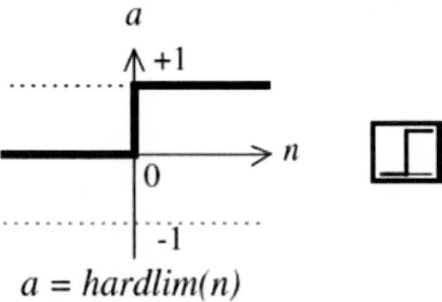

Figure 2. Hard-Limit transfer function.

Three commonly used functions are "hard limit", "linear" and "sigmoid" [10]. The hard-limit transfer function shown in Figure 2 limits the output of the neuron to either 0, if the net input argument n is less than 0, or 1, if n is greater than or equal to 0. More specifically, a negative input does not pass the threshold, then the function returns the value 0 (0 can be interpreted as false), then a nonnegative input exceeds the threshold, and the function returns 1 (true).

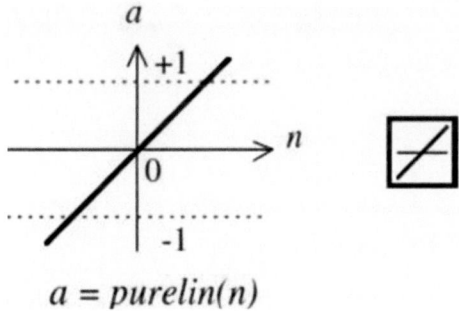

Figure 3. Linear transfer function.

It must be noticed that the bias b in the equation $a = hardlim\ (w^T p - b)$ determines the location of the axis $w^T p$, where function goes from 0 to 1.

The other important transfer function is called linear and shown below (see Figure 3).

The linear function is very simple; it directly transfers its input to its output:

$$a = n \qquad (1)$$

In this case, the output of the neuron corresponds to its level of activation and the zero output occurs when:

$$w^T p = b \qquad (2)$$

The sigmoid transfer function shown in Figure 4 takes the input, which can have any value between plus and minus infinity, and squashes the output into the range 0 to 1.

Most widely used transfer function for the input and hidden layers is the sigmoid transfer function and given by Eq. (3).

$$a = \frac{1}{1+e^{-n}} \qquad (3)$$

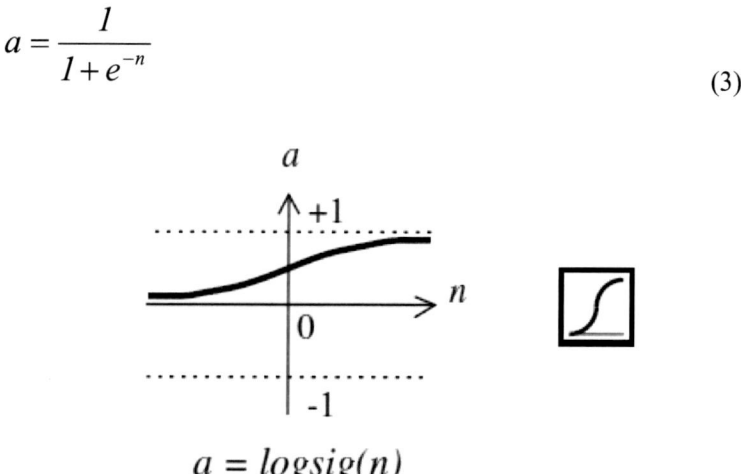

Figure 4. Sigmoid transfer function.

The sigmoid function is very non-linear because there is a discontinuity when $w^T p = b$. This transfer function is commonly used in back propagation networks, because it is differentiable. Finally, it must be noted that the "hyperbolic tangent" is a symmetrical version of the sigmoid function [10].

2.2. LEARNING PROCESS

Among the desirable properties of a neural network, the ability of learning from its environment to improve its performance is surely the most important. But what then is learning? Unfortunately, there is no general, universally accepted definition, as this term is associated with many different concepts that depend on each person's perspective. Learning is a dynamic and iterative process that modifies the parameters of a network. This process is a respond to the signals that the network receives from its environment [11].

In most topologies, learning results to a change of synaptic efficiency, or in other words, to a change in the amount of the weights connecting the neurons of one layer to another. If the weight $w_{i,j}$ connects neuron i to entry j, a change of weight $\Delta w_{i,j}(t)$ at time t, can be expressed simply as follows:

$$\Delta w_{i,j}(t) = w_{i,j}(t+1) - w_{i,j}(t) \qquad (4)$$

and, therefore,

$$w_{i,j}(t+1) = w_{i,j}(t) + \Delta w_{i,j}(t) \qquad (5)$$

where $w_{i,j}(t+1)$ and $w_{i,j}(t)$ represent the new and old amounts of weight $w_{i,j}$, respectively. A set of well-defined rules to achieve such an adaptation process of weight is called the learning algorithm of the neural network [12].

The word "algorithm" comes from the name of a Persian mathematician Muhammed ibn Musa al-Khwarizmi, who lived in the 9th century A.D. In Latin, its name was translated by Algorismus, which later turned into an algorithm [10].

2.2.1. Supervised Learning

Supervised learning is characterized by the presence of a "professor" who has a deep knowledge of the environment in which the neural network evolves. In practice, knowledge of the professor is in the form of Q pairs of input/output data sets which we denote $((p_1, d_1), (p_2, d_2). . ., (p_Q, d_Q))$, where p_i refers to an input and d_i is the desired output of the network. Each pair (p_i, d_i) corresponds to a case that the network should produce the target for a given input. For this reason, supervised learning has been also described as learning by examples [13].

Supervised learning is illustrated schematically in Figure 5. The environment is unknown to the network. This produces an input $p(t)$ that is routed to both the professor and the network. Thanks to its intrinsic knowledge, the professor produces a desired output $d(t)$ for this input. It is assumed that this response is optimal. It is then compared (by subtraction) with the output of the network to produce an error signal $e(t)$ which will be re-injected into the network to change its behavior through an iterative process. This iteration eventually allows the network to simulate the response of the professor. In other words, knowledge of the environment by the professor is gradually transferred to the network until it reaches a certain stopping criterion. Subsequently, we can eliminate the professor and let the network operate independently [13].

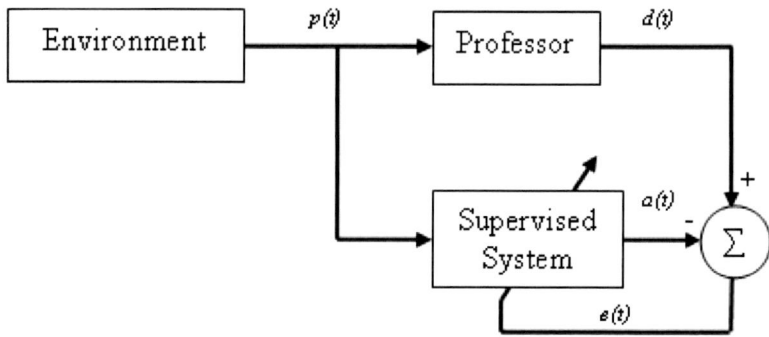

Figure 5. Supervised learning scheme.

2.2.2. Reinforcement Learning

The reinforcement learning overcomes some limitations of supervised learning. It is a kind of supervised learning, but with a hint of satisfaction instead of a scalar error signal vector. In practice, the use of reinforcement learning is complex to implement. But, it is important to understand the difference between this type of learning and supervised one. Supervised learning has an error signal that not only calculates an index of satisfaction (e.g. mean square error), but also estimates the local gradient indicating a direction for the adaptation of synaptic weights. This information is provided by the professor who makes all the difference. In reinforcement learning, the absence of error signal makes the calculation of this gradient impossible. To estimate the gradient, the network is forced to attempt actions and observe the outcome, eventually infer a direction of change for the synaptic weights. To do this, it implements a process of trial/error while delaying the reward offered by the satisfaction index [14].

Thus, we introduce two distinct phases: an exploration phase where the network tries random directions of change, and an exploitation phase where it makes a decision. This two-step process can significantly slow learning. In addition, it introduces a dilemma between the desire to use of the information already learned about the merits of different actions, and that of new knowledge about the consequences of those decisions [14].

2.2.3. Unsupervised Learning

This form of learning, called "unsupervised" or "self-organized", is characterized by the complete absence of professor, neither a signal error, as in the case of supervised, nor an index of satisfaction, as in the case for reinforcement. Therefore, there is an environment that provides inputs, and a network must learn without external intervention. By assimilating the input from the environment to a description of its internal state, the major task of a network is then to model this state as best as possible. To achieve this, we must first define a measure of quality for this model and use it later to optimize free parameters in the network, i.e. its synaptic weights. At the end of learning process, the network has developed an ability to form internal representations of environmental variations permitting to encode their characteristics and, therefore, automatically create similar classes of responses [15].

The unsupervised learning is typically based on a competitive process to create a model where the synaptic weights of neurons represent corresponding targets of inputs. The quality of the resulting model must be estimated using a method to measure the distance between inputs and their targets [15].

2.3. TRAINING ALGORITHMS

2.3.1. Back Propagation Algorithm

Back propagation was created by generalizing the Widrow-Hoff learning rule to multiple-layer networks and nonlinear differentiable transfer functions. Input vectors and the corresponding target vectors are used to train a network until it can approximate a function, associate input vectors with specific output vectors, or classify input vectors in an appropriate way as defined by you [16].

Standard back propagation is a gradient descent algorithm in which the network weights are moved along the negative of the gradient of the performance function. The term back propagation refers to the manner in which the gradient is computed for nonlinear multilayer networks. There are a number of variations on the basic algorithm that are based on other standard optimization techniques, such as conjugate gradient and Newton methods.

Properly trained back propagation networks tend to give reasonable answers when presented with inputs that they have never seen. Typically, a new input leads to an output similar to the correct output for input vectors used in training that are similar to the new input being presented. This generalization property makes it possible to train a network on a representative set of input/target pairs and get good results without training the network on all possible input/output pairs.

There are many variations of the back propagation algorithm, several of which are described in this chapter. The simplest implementation of back propagation learning updates the network weights and biases in the direction in which the performance function decreases most rapidly, the negative of the gradient. An iteration of this algorithm can be written as follows:

$$x_{k+1} = x_k - \alpha_k g_k \tag{6}$$

where x_k is a vector of current weights and biases, g_k is the current gradient, and α_k is the learning rate [16].

There are two different ways in which gradient descent algorithm can be implemented: incremental mode and batch mode. In incremental mode, the gradient is computed and the weights are updated after each input is applied to the network. In other words the network is presented with cases from the training data one at a time and the weights are revised after each case in an attempt to minimize the error function. In batch mode, all the inputs are applied to the network before the weights are updated [17]. Some important types of back propagation algorithms are described below.

2.3.1.1. Conjugate Gradient Algorithm

The basic back propagation algorithm adjusts the weights in the steepest descent direction (negative of the gradient); the direction in which the performance functions is decreasing most rapidly. It turns out that, although the function decreases most rapidly along the negative of the gradient, this does not necessarily produce the fastest convergence. In the conjugate gradient algorithms a search is performed along conjugate directions, which produces generally faster convergence than steepest descent directions [18].

The gradient method has most of the benefits of Newton's method but without the inconvenience of having to calculate and invert the Hessian matrix. The Hessian matrix (or simply the Hessian) is the square matrix of second-order partial derivatives of a function; it describes the local curvature of a function of many variables [19].

It is based on the concept of conjugate vectors, and on other hand, finding a minimum along a line. The vectors of a set (p_k) are mutually conjugate with respect to a positive definite matrix A (whose values are all strictly positive) if and only if:

$$p_k^T A p_j = 0 \ , \ k \neq j \qquad (7)$$

As with orthogonal vectors, there are an infinite number of sets of vectors which covers a combined vector space of dimension of m. One of those is made up of vectors A, $(z_1, z_2, \ldots, z_m)$, associated with proper values $(\lambda_1, \lambda_2, \ldots, \lambda_m)$. To show this, it is simply sufficient to replace p_k with z_k in the previous equation:

$$z_k^T A z_j = z_k^T \lambda_j z_j = \lambda_j z_k^T z_j = 0 \ , \ k \neq j \qquad (8)$$

where this equality shows the fact that the vectors of a positive definite matrix are always orthogonal [13]. Therefore, the vectors of such a matrix are both orthogonal and conjugate. For a quadratic function F with m free parameters, we can show that it is always possible to reach a minimum by performing at most m linear search along lines oriented in conjugate direction $(p_1, p_2, \ldots, p_m)$. The question that remains is how to build these directions together without reference to the Hessian matrix of F?

The general expression for a quadratic function is given by:

$$F(x) = \frac{1}{2} x^T A x + d^T x + c \qquad (9)$$

where the gradient ∇F is given by:

$$\nabla F(x) = Ax + d \qquad (10)$$

and the Hessian matrix by $\nabla 2F(x) = A$.

Putting $g_t \equiv \nabla F(x)|_{x=xt}$ and combining these two equations, we can find the change of gradient Δg_t at iteration t:

$$\Delta g_t = g_{t+1} - g_t = (Ax_{t+1} + d) - (Ax_t + d) = A\Delta x_t \quad (11)$$

where the variation of parameters Δx_t at time t is given by:

$$\Delta x_t = x_{t+1} - x_t = \alpha_t p_t \quad (12)$$

with α_t chosen to minimize $F(x_t)$ in the direction of p_t.

We can now rewrite the condition of conjugate vectors as follows:

$$\alpha_t p_t^T A p_j = \Delta x_t^T A p_j = \Delta g_t^T p_j = 0 \quad t \neq j \quad (13)$$

It can be noticed that considering the change in gradient at each iteration *(t)* of the algorithm, the Hessian matrix of the equation that defines the condition of conjugate vectors can be removed. The search direction p_j will be combined provided they are orthogonal to the gradient variation [13].

At each iteration *(t)* of the conjugate gradient algorithm, it therefore builds a search direction that is orthogonal to p_t (Δg_0, Δg_1, ..., Δg_{t-1}) using a procedure similar to the method of Gram-Schmidt that can be simplified to the following expression:

$$p_t = -g_t + \beta_t p_{t-1} \quad (14)$$

where scalars β_t can be calculated in three equivalent ways:

$$\beta_t = \frac{\Delta g_{t-1}^T g_t}{\Delta g_{t-1}^T p_{t-1}}, \quad \beta_t = \frac{g_t^T g_t}{\Delta g_{t-1}^T g_{t-1}}, \quad \beta_t = \frac{\Delta g_{t-1}^T g_t}{\Delta g_{t-1}^T g_{t-1}} \quad (15)$$

2.3.1.2. Scaled Conjugate Gradient Algorithm

Conjugate gradient algorithm discussed above requires a line search at each iteration. This line search is computationally expensive, because it requires that the network response to all training inputs be computed several times for each search. The scaled conjugate gradient algorithm (*scg*), developed by Moller [11], and was designed to avoid the time-consuming line search. This algorithm combines the model-trust region approach, used in the Levenberg-Marquardt algorithm, described in section 2.4.3.

2.3.2. Quasi-Newton Algorithms

Newton's method is an alternative to the conjugate gradient methods for fast optimization. The basic step of Newton's method is:

$$x_{k+1} = x_k - A_k^{-1} g_k \qquad (16)$$

where A_k^{-1} is the Hessian matrix (second derivatives) of the performance index at the current values of the weights and biases. Newton's method often converges faster than conjugate gradient methods. Unfortunately, it is complex and expensive to compute the Hessian matrix for feed-forward neural networks.

There is a class of algorithms that is based on Newton's method, but which doesn't require calculation of second derivatives. These are called quasi-Newton (or secant) methods. They update an approximate Hessian matrix at each iteration of the algorithm. The update is computed as a function of the gradient [10, 20].

2.3.3. Levenberg-Marquardt Algorithm

Like the quasi-Newton methods, the Levenberg-Marquardt algorithm was designed to approach second-order training speed without having to compute the Hessian matrix. When the performance function has the form of a sum of squares (as is typical in training feed-forward networks), then the Hessian matrix can be approximated as:

$$H = J^T J \qquad (17)$$

and the gradient can be computed as:

$$g = J^T e \qquad (18)$$

where J is the Jacobian matrix that contains first derivatives of the network errors with respect to the weights and biases, and e is a vector of network errors. The Jacobian matrix can be computed through a standard back propagation technique that is much less complex than computing the Hessian matrix.

The Levenberg-Marquardt algorithm uses this approximation to the Hessian matrix in the following Newton-like update:

$$x_{k+1} = x_k - [J^T J + \mu I]^{-1} J^T e \qquad (19)$$

When the scalar μ is zero, this is just Newton's method, using the approximate Hessian matrix. When μ is large, this becomes gradient descent with a small step size. Newton's method is faster and more accurate near an error minimum, so the aim is to shift toward Newton's method as quickly as possible [20].

Thus, μ is decreased after each successful step (reduction in performance function) and is increased only when a tentative step would increase the performance function. In this way, the performance function is always reduced at each iteration of the algorithm.

Chapter 3

TRAINING, VALIDATION AND TEST OF A NEURAL NETWORK

The primary objective of this chapter is to explain how to use the training functions to train neural networks to solve specific problems. There are generally four steps in the training process:

- Assemble the training data,
- Create the network object,
- Train the network, and
- Simulate the network response to new inputs.

This chapter discusses a general optimization process of a neural network topology. It describes the architecture of a typical multilayer feed-forward network. Then the simulation and training of the network objects will be presented.

To describe such a network, a brief description of an elementary neuron with R inputs is shown in Figure 6. Each input is weighted with an appropriate w. The sum of the weighted inputs and the bias forms the input to the transfer function f. Neurons can use any differentiable transfer function f (Presented in Table 1) to generate their output.

Networks with biases, a sigmoid layer, and a linear output layer are capable of approximating any function with a finite number of discontinuities

Feed-forward networks often have one or more hidden layers of sigmoid neurons followed by an output layer of linear neurons. Multiple layers of neurons with nonlinear transfer functions allow the network to learn nonlinear and linear relationships between input and output vectors. The linear output layer lets the network produce values outside the range -1 to +1.

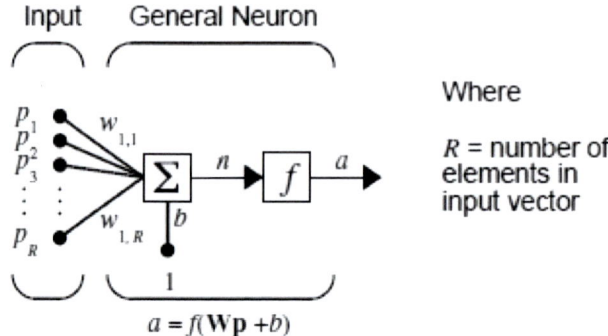

Figure 6. An elementary neuron with R inputs.

On the other hand, if the outputs of a network must be constrained (such as between 0 and 1), the output layer should use a sigmoid transfer function (such as *logsig*).

In order to determine the optimum number of hidden nodes, a series of topologies are used, in which the number of nodes was varied from for example 2 to 20. A mathematic function such as the mean square error (MSE) is normally used as the error function. MSE measures the performance of the network according to the following equation:

$$MSE = \frac{\sum_{i=1}^{i=N}(y_{i,pred} - y_{i,\exp})^2}{N} \qquad (20)$$

where N is the number of data point, $y_{i,pred}$ is the network prediction, $y_{i,exp}$ is the experimental response and i is an index of data.

All ANNs are trained using one appropriate algorithm such as scaled conjugate gradient, quasi-Newton, Levenberg-Marquardt algorithms and so on. Network training is a process by which the connection weight and bias on the ANN are adapted through a continuous process of simulation by the environment in which the network is embedded. The primary goal of training is minimizing the error function by searching for a set of connection weights and biases that causes the ANN to produce outputs that are equal or close to target values. For instance, the back propagation algorithm minimizes the error function between the observed and the predicted output in the output layer, through two phases (Figure 7). In the forward phase, the external input information signals at the input neurons are propagated forward to compute the output information signal at the output neuron. In the backward phase, modifications to the connection strengths are made based on the basis of the difference in the predicted and observed information signals at the output neuron [21].

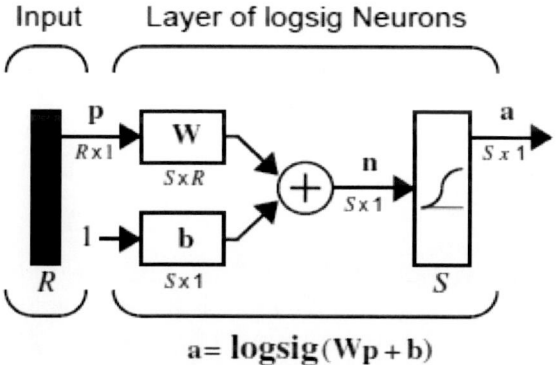

Figure 7. A layer diagram of feed-forward network of S *logsig* neurons having R inputs.

If the used transfer function in the hidden layer is sigmoid, all samples must be normalized in the range of 0.1-0.9 [5]. So all of the data sets (X_i) were scaled to a new value x_i as follows:

$$x_i = 0.8 \left(\frac{X_i - X_{min}}{X_{max} - X_{min}} \right) + 0.1 \tag{21}$$

Each topology would be repeated several times to avoid random correlation due to the random initialization of the weights. Figure 8 illustrates a typical graph of the network error versus the number of neurons in the hidden layer. It could be seen that the network performance stabilized after inclusion of defined number of nodes on the hidden layer.

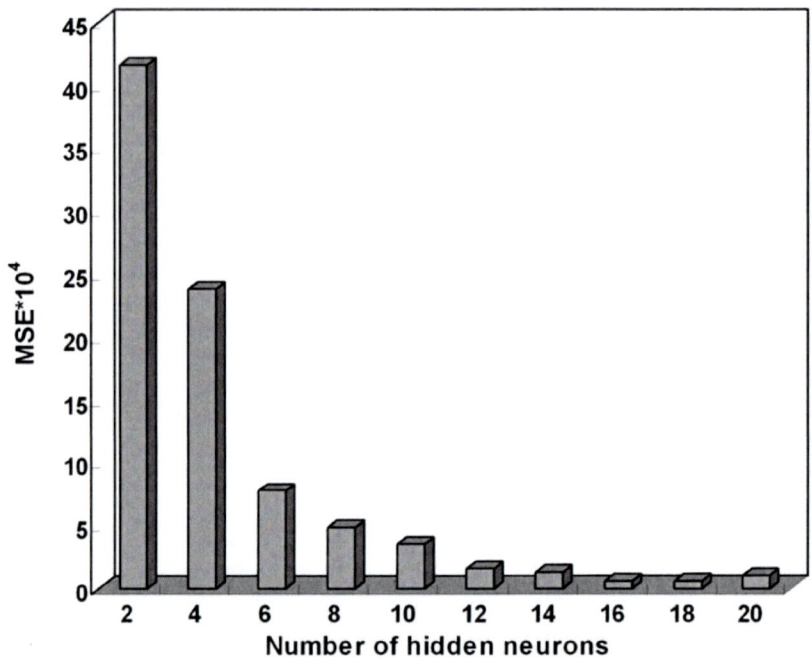

Figure 8. Effect of the number of neurons of hidden layer on the performance of the neural network. (Feed-forward back propagation network trained with scaled conjugate gradient algorithm; topology 4:16:1; data number 117) [22].

Salari et al. [22] have shown that based on the approximation of error function (MSE), a number of hidden neurons equal to sixteen was adopted in a feed-forward back propagation neural network and was used for modeling of the process studied.

3.1. TEST OF THE FITTED MODEL

In order to calculate modeling errors, all of the outputs would performed an inverse range scaling to return the predicted responses to their original scale and compared them with experimental responses. Figure 9 shows a typical comparison between experimental values and predicted output variables using adopted neural network model.

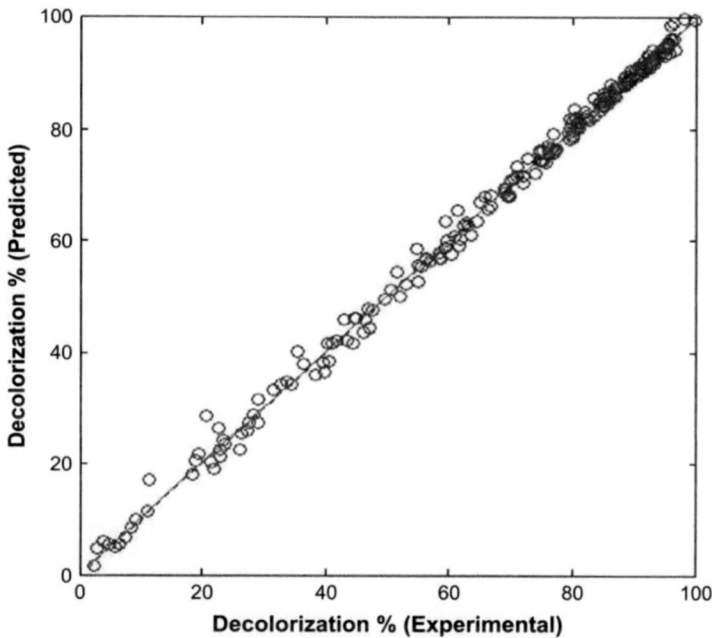

Figure 9. Comparison between experimental values and predicted output variables using adopted neural network model; $R^2 = 0.996$ (Feed-forward back propagation network trained with scaled conjugate gradient algorithm; topology 4:8:1; data number 228) [23].

Usually two lines are used to show the success of the prediction [23, 24]. The one is the perfect fit (predicted data = experimental data), on which all the data of an ideal model should lay. The other line is the line that best fits on the data of the scatter plot with equation $Y = ax + b$ and it is obtained with regression analysis based on the minimization of the squared errors. If there were a perfect fit (outputs exactly equal to targets), the slope would be 1, and the y-intercept would be 0. In Figure 9, it could be seen that these values are very close to 1 and 0. The third variable is the correlation coefficient (R-value) between the outputs and targets. It is a measure of how well the variation in the output is explained by the targets. If this number is equal to 1, then there is perfect correlation between targets and outputs. In this example, the correlation coefficient is 0.996, which indicates a good fit. This figure confirms that the neural network model could effectively reproduce the experimental results.

3.2. Relative Importance of Input Variables

The ANN used provides the weights that are coefficients between the artificial neurons. These weights are analogous to synapse strengths between the axons and dendrites in real biological neurons. Therefore, each weight decides what proportion of the incoming signal will be transmitted into the neuron's body [25].

The neural net weight matrix can be used to assess the relative importance of the various input variables on the output variables. Garson [26] has proposed an equation based on partitioning of connection weights:

$$I_j = \frac{\sum_{m=1}^{m=N_h}\left(\left(\left|W_{jm}^{ih}\right|\Big/\sum_{k=1}^{N_i}\left|W_{km}^{ih}\right|\right)\times\left|W_{mn}^{ho}\right|\right)}{\sum_{k=1}^{k=N_i}\left\{\sum_{m=1}^{m=N_h}\left(\left|W_{km}^{ih}\right|\Big/\sum_{k=1}^{N_i}\left|W_{km}^{ih}\right|\right)\times\left|W_{mn}^{ho}\right|\right\}} \quad (22)$$

where I_j is the relative importance of the jth input variable on output variable, N_i and N_h are the number of input and hidden-neurons, respectively, W's are connection weights, the superscripts 'i', 'h' and 'o' refer to input, hidden and output layers, respectively, and subscripts 'k', 'm' and 'n' refer to input, hidden and output neurons, respectively.

Kasiri et al. [24] have used this equation to evaluate the relative importance of the input variables on color removal of the dye solution by UV/H_2O_2 process.

3.3. Improving Generalization

One of the problems that occur during neural network training is called "overfitting". The error on the training set is driven to a very small value, but when new data is presented to the network the error is large. The network has memorized the training examples, but it has not learned to generalize to new situations.

Figure 10 shows the response of a 1-20-1 neural network that has been trained to approximate a noisy sine function. The underlying sine function is shown by the dotted line, the noisy measurements are given by the '+'

symbols, and the neural network response is given by the solid line. Clearly this network has overfitted the data and will not generalize well [10].

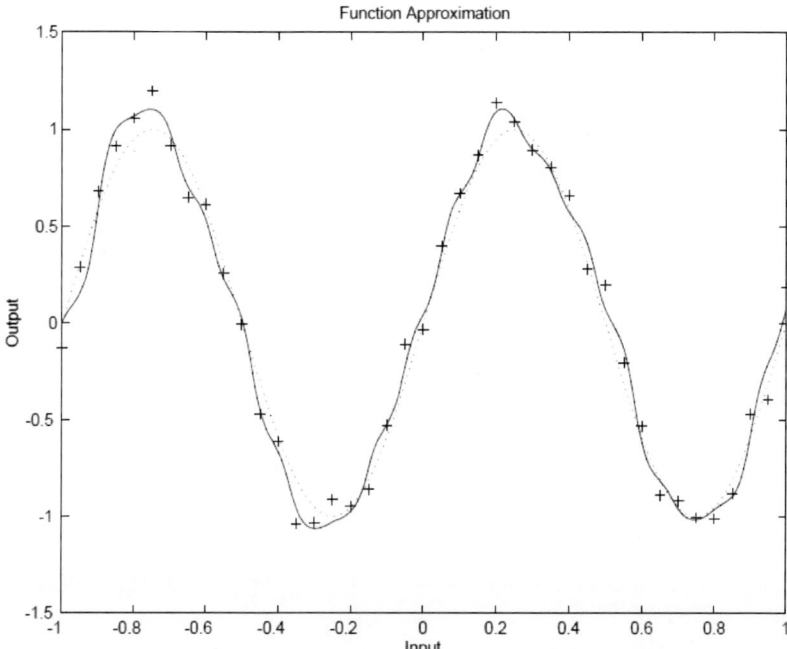

Figure 10. A typical overfitted neural network.

One method for improving network generalization is to use a network that is just large enough to provide an adequate fit. The larger network you use, the more complex the functions the network can create. If you use a small enough network, it will not have enough power to overfit the data. Unfortunately, it is difficult to know beforehand how large a network should be for a specific application. There are two other methods for improving called "regularization" and "early stopping". The next sections describe these two techniques and the routines to implement them. Note that if the number of parameters in the network is much smaller than the total number of points in the training set, then there is little or no chance of overfitting. If you can easily collect more data and increase the size of the training set, then there is no need to worry about the following techniques to prevent overfitting. The rest of this section only applies to those situations in which you want to make the most of a limited supply of data [12].

3.3.1. Regularization

The first method for improving generalization is called regularization. This involves modifying the performance function, which is normally chosen to be the sum of squares of the network errors on the training set.

For instance in Bayesian regularization, the weights and biases of the network are assumed to be random variables with specified distributions. The regularization parameters are related to the unknown variances associated with these distributions, and these parameters can be estimated using statistical techniques [12].

3.3.2. Early Stopping

Another method for improving generalization is called early stopping. In this technique the available data is divided into three subsets. The first subset is the training set, which is used for computing the gradient and updating the network weights and biases. The second subset is the validation set. The error on the validation set is monitored during the training process. The validation error normally decreases during the initial phase of training, as does the training set error. However, when the network begins to overfit the data, the error on the validation set typically begins to rise. When the validation error increases for a specified number of iterations, the training is stopped, and the weights and biases at the minimum of the validation error are returned.

The test set error is not used during the training, but it is used to compare different models. It is also useful to plot the test set error during the training process. If the error in the test set reaches a minimum at a significantly different iteration number than the validation set error, this might indicate a poor division of the data set.

The data sets would be divided into training, validation and test subsets that each of them normally contains 50, 25 and 25% of samples, respectively. The validation and test sets, for evaluation of the validation and modeling power of the nets, must be randomly selected from the experimental data.

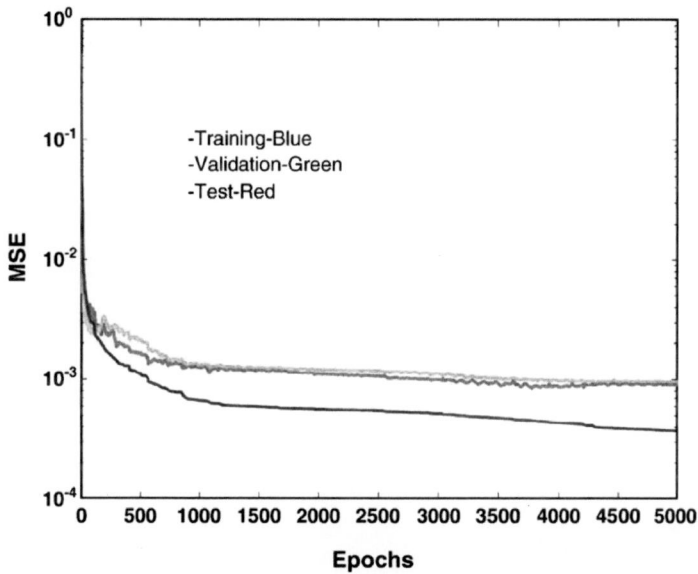

Figure 11. The progress of the mean square error (MSE) with the number of iterations of training, validation and test. (Feed-forward back propagation network trained with scaled conjugate gradient algorithm; topology 4:8:1; data number 177) [27].

Khataee and Mirzanjani [27] have described that a good method for preventing the overfitting is to use the validation data set periodically to compute the error rate for it while the network is being trained. The validation error decreases in the early epochs of back propagation but after a while, it begins to increase. The point of minimum validation error is a good indicator of the best number of epochs for training and the weight at that stage are likely to provide the best error rate in new data [5]. Their results indicated that the minimum error of the validation set could be achieved in the epochs just about 5000. After 5000 epochs the mean square error slightly increased; therefore, 5000 has been selected as the optimum epoch number. Figure 11 shows the progress of the mean square error with the number of iterations of training, validation and test.

Chapter 4

APPLICATIONS OF ARTIFICIAL NEURAL NETWORK MODELING

Neural network technology has been developed in an attempt to mimic the acquisition of knowledge and organization skills of the human brain. It offers significant support in terms of organizing, classifying, and summarizing data. It also helps to discern patterns among input data, requires few assumptions, and achieves a high degree of prediction accuracy. These characteristics make neural network technology a potentially promising tool for recognition, classification, prediction, and optimization in various areas, ranging from finance to medicine [28, 29]. Applications of ANNs in different areas with selected examples have been reported in Table 2.

Financial application areas that require pattern matching, classification, and prediction, such as bankruptcy prediction, loan evaluation, credit scoring, and bond rating, are fruitful candidate areas for neural network technology. Great strides have been made in this technology in the last decade. This has led to increasing efforts to use it in a wide variety of scientific applications and has contributed to the development of many different types of business applications (see Table 2). In the literature, an increasing amount of information has appeared, with a considerable portion focusing on the actual neural network development in the area of finance [30-33]. The historical trends in published business and financial ANNs application research have been reviewed by Wong and co-workers [28, 34]. Much of the research on the business and financial ANNs applications have been focused on bankruptcy prediction, credit evaluation, insolvency prediction, fraud detection and property evaluation [35]. On the other hand, the application of ANN in marketing area is relatively new but is becoming popular because of their ability of capturing nonlinear relationship between the variables. Numerous applications of ANNs models in marketing discipline are available, to mention a few are market segmentation, market response prediction, new product launch, sales forecasting and consumer choice prediction [35-37] (see Table 2).

Although ANNs applications in various aspects of engineering have been reported, there have been a wide range of its applications in areas relevant to chemical engineering recently (see Table 2). This can be attributed to many reasons, some of which are as follows [38]:

- The tremendous hardware advances in digital technology have enabled simulations of neural nets economically and with relative ease and speed;
- Application of ANNs for sensor pattern classification has been found to be superior to the traditional algorithmic techniques or the expert system approaches;
- ANNs offer the promise of being able to extract information from plant in an efficient manner with normal availability of rich data. In some cases, it may not be cost effective to develop models from first principles at all times, especially those dealing with severe/unknown non-linearity which commonly found in chemical process systems. ANNs are the simple and efficient alternatives;
- Use of ANNs in the middle ground between mathematical-based and black box approaches for solving many classes of problems.

In the last decade, the use of artificial neural networks has become widely accepted in medical applications (see Table 2). This is manifested by an increasing number of medical devices currently available on the market with embedded ANN algorithms. Claimed advantages of neural network applications in medicine include [39]:

- Ease of optimization, resulting in cost-effective and flexible non-linear modeling of large data sets;
- Accuracy for predictive inference, with potential to support clinical decision making;
- These models can make knowledge dissemination easier by providing explanation, for instance, using rule extraction or sensitivity analysis.

Numerous articles involving applications of ANNs in medicine have been published over the years [39-41]. Baxt has presented applications of neural networks in clinical medicine. The review illustrates the huge flexibility of the ANN paradigm and its ability, in a wide variety of areas, to perform with significant diagnostic accuracy [42]. Reggia [43] has briefly explained the nature of a neural model and then reviews the work in neural computation involving problems in medical informatics (e.g. expert systems) and modeling of psychiatric and neurological phenomena.

ANNs have been also used to predict the grain size of different nanocrystals (see Table 2) [44, 45]. For instance, a feed-forward multilayer perceptron artificial neural network framework has been used to model the dependence of the grain size of nano-crystalline nickel coatings on the process parameters namely current density, saccharin concentration and bath temperature. The process parameters are used as the model inputs and the resulting grain size of the nano-crystalline coating as the output of the model. Comparison between the model predictions and the experimental observations indicate a remarkable agreement between them [45]. During the following chapters, application of neural network modeling in different processes of water and wastewater treatment will be discussed in details.

Table 2. Applications of artificial neural networks in different areas

Application area	Remarks	Refs.
Finance	Bankruptcy prediction, loan evaluation, credit scoring, bond rating, checking account overdrafts, future options hedging and pricing, real estate appraisal, and interest rate prediction	[28, 34]
Accounting	Audit opinion prediction, auditor's going concern uncertainty decision, tax form processing, and litigation prediction	[46-47]
Marketing	Consumer segments identification, future order forecasting, market response functions determination, market responses prediction, product purchase frequency prediction, solo mailing, and telephone interview response analysis	[48-51]
Human resources	Personnel selection, salespeople hiring, and workplace behavior prediction	[34]
Information systems	Computer access security, computer program risk analysis, computer users authentication, computer users identification, computer viruses recognition and classification, database tables clustering, software development effort, and pictorial information retrieval	[52-55]

Table 2. (Continued).

Application area	Remarks	Refs.
Production/ Operations	Production scheduling, automated food inspection, automated guided vehicle system optimization, beam landing adjustment, cellular manufacturing systems design, economic power dispatch, intelligent packaging, and job scheduling	[56, 57]
Image processing	Image and video compression, image reconstruction, image restoration, image enhancement, image compression, feature extraction, segmentation, object recognition, and image understanding	[58-60]
Control strategies	Internal model controller, model predictive, extended dynamic matrix control, generic model control, generalized predictive controller, and feed-back controller	[38, 61]
Health and medicine	Sleep research; diagnosis, prognosis and image analysis in cancer; clinical trials and randomized control trials with ANN in cardiology, oncology, neurology, critical care and prostatic, and cervical and breast cancer	[15, 40, 41, 62-64]
Chemical engineering	Dynamic modeling of chemical process, design of multi-component catalysts, refinery fluid catalytic cracking, automated control, multivariate calibration, modeling of cracking of Naphtha, foodstuff analyses, prediction the properties of polymer composites, sensor pattern classification, and modeling of water & wastewater treatment processes	[38, 56, 65, 66]
Environmental systems	Modeling and simulation of export of nutrients from river basins, forecast salinity, algal growth and transport in rivers, ozone levels, classifying soil structure, prediction of air pollution, and the characteristics of ecosystems	[67-72]
Nanotechnology and nanoscience	Prediction the grain size of nickel, Cd–Mn–S and TiO_2 nanocrystals, prediction dynamic hysteresis loops of nano-crystalline cores, and purity of the perovskite-type $SrTiO_3$ nanocrystals	[44, 73-75]
Defense	Weapon steering, target tracking, object discrimination, facial recognition, new kinds of sensors, sonar, radar and image signal processing including data compression, feature extraction and noise suppression, and signal/image identification	[10, 76-78]
Transportation	Truck brake diagnosis systems, vehicle scheduling and routing systems, estimation of freeway travel time, modeling public transport trips	[10, 79-81]
Aerospace	High performance aircraft autopilot, flight path simulation, aircraft control systems, autopilot enhancements, aircraft component simulation, and aircraft component fault detection	[10, 82-85]

Chapter 5

ANN MODELING OF ADSORPTION PROCESSES

The adsorption process is very effective in removing dissolved matter from water and wastewater. Adsorption has been found to be superior to other techniques for water reuse in terms of initial cost, flexibility and simplicity of design, ease of operation and insensitivity to toxic pollutants. Adsorption process also does not result in the formation of harmful substances [86, 87].

The adsorption processes are effective in the removal of various organic substances such as oils, radioactive compounds, petroleum hydrocarbons, poly aromatic hydrocarbons, pesticides, textile dyes and various halogenated compounds like chlorine, fluorine, bromine and iodine. Apart from organic compounds, the adsorption processes also remove inorganic compounds like arsenic, cadmium, chromium, zinc, lead, mercury, copper, etc. [86-90].

Activated carbon materials have a special place among the adsorbents, as for a long time they are known to be capable of adsorbing various organic compounds. On the basis of physical properties, activated carbon can be classified as powdered activated carbon (PAC), granulated activated carbon (GAC), pelleted activated carbon, spherical activated carbon, and impregnated carbon.

Although activated carbon is the most widely used for the removal of a variety of organics from waters, it has some disadvantages like the high regeneration cost and the generation of carbons fines. Thus, some natural and synthetic adsorbents have been recently used for removal of various soluble substances from water. Attention has been focused on various adsorbents, which have absorption capacities and are able to remove unwanted substances from contaminated water at low-cost. Because of their low-cost and local availability, natural materials such as bagasse, red mud, zeolites, clay, or certain waste products from industrial operations such as agricultural and industrial wastes, fly ash, and coal are classified as low-cost adsorbents [86, 87].

Artificial neural networks have been used for modeling of water and wastewater treatment in the presence of various adsorbents (see Table 3). For instance, the three-layered feed-forward back propagation neural network has been used for modeling of nitrate adsorption on granular activated carbon [91]. It should be noted that high concentrations of N-containing compounds in drinking water cause health problems such as cyanosis among children and cancer of the alimentary canal. The input variables to the feed-forward neural network are the amount of GAC (g), initial concentration of nitrate (ppm), contact time (min), initial pH and temperature (°C). The percentage of nitrate removal has been chosen as the experimental response or output variable. The sigmoidal transfer function has been used as a transfer function in the hidden and output layers. The training function is Scaled Conjugate Gradient (*trainscg*). Totally 468 experimental sets have been used to develop the ANN model. The linear regression between the network prediction and the corresponding experimental data (R^2=0.997) proves that modeling of nitrate adsorption process using artificial neuron network is a good and precise method to predict the extent of adsorption of nitrate on GAC under different conditions [91].

Aber *et al.* [92] have reported the modeling of acid orange 7 removal by powdered activated carbon from aqueous solutions with the initial dye concentrations of 150 ppm to 350 ppm and initial pH values of 2.8, 5.8, 8.0 and 10.5 at 25 °C. ANN modeling of experimental results has been performed using a 3-layer feed-forward back propagation network with 3, 2 and 1 neurons in first, second and third layers, respectively. Transfer function of neurons at first and second layers was tan-sigmoid and at output layer was log-sigmoid. Training function is *trainscg*. Total 219 experimental points have been randomly split between training and prediction sets with 2:1 ratio, respectively. The mean relative error of modeling was 5.81%.

A three-layer artificial neural network model has been developed to predict the efficiency of Pb(II) ions removal from aqueous solution by Antep pistachio (*Pistacia Vera* L.) shells based on 66 experimental sets. The trained ANN model with Levenberg–Marquardt algorithm has been able to predict adsorption efficiency with a tangent sigmoid transfer function (*tansig*) at hidden layer with 11 neurons and a linear transfer function (*purelin*) at output layer. The linear regression between the network outputs and the corresponding targets is proven to be satisfactory with a correlation coefficient of about 0.936 [93].

The ANN modeling of three pesticides adsorption onto five different activated carbons with different characteristics in terms of shapes, dimensions and pore properties has been also reported. The selection of data is based on the physical and statistical approaches, equilibrium parameters assessed in static reactors being considered as influential variables to take into account the competitive adsorption phenomenon. Static and recurrent neural networks provide both high correlation coefficient ($R^2 > 0.990$) and low root mean square modeling errors (RMSE < 0.035 while standard deviation of data is equal to 2.9%) for the prediction of the global breakthrough curves [94].

Adsorption of bovine serum albumin on porous polyethylene membrane has been modeled using a back propagation artificial neural network with two inputs (nondimensionalized depth and adsorption time) and one output (nondimensionalized concentration). The data pairs are divided into training and validating sets. The validating set has the correlation coefficient $R^2 = 0.9888$ and RMSE = 0.0082, and the training set has the correlation coefficient $R^2 = 0.9779$ and RMSE = 0.0098 [95].

Table 3. ANN modeling of adsorption processes for water and wastewater treatment

Adsorbent	Removal Target	ANN Architecture	Training Function	Layers No.	ANN Topology	Data No.	Input	Output	Epochs No.	Ref.
Granular activated carbon (GAC)	Nitrate	Feed-forward back propagation	Train scaled conjugate gradient	3	5:8:1	468	$[GAC]_0$, $[Nitrate]_0$, Time, pH, Temperature	Nitrate removal (%)	–	[91]
Powdered activated carbon	Acid Orange 7	Feed-forward back propagation	Train scaled conjugate gradient	3	3:2:1	219	$[Dye]_0$, Time, pH	$[Dye]_t$ (mg/L)	–	[92]
Antep pistachio (Pistacia Vera L.) shells	Pb(II)	Feed-forward back propagation	Levenberg–Marquardt algorithm	3	5:11:1	66	$[Adsorbent]_0$, $[Pb(II)]_0$, pH, Time, Temperature	Pb(II) removal (%)	100	[93]
Activated carbon filters	Three pesticides: atrazin, atrazin-desethyl and triflusulfuron-methyl	Feed-forward back propagation	Levenberg–Marquardt algorithm	3	13:6:1	9749	*Binary variable, Microspores vol., Mesoporous vol., Solubility, Mw, C_0, TOC_0, Flow velocity, K_1, $1/n_1$, K_2, $[TiO_2]_0$, NOM Elimination(%), Time	$[pesticide]_t/[pesticide]_0$	100	[94]

* **NOM**: Natural Organic Matter; n_i: Freundlich parameter of compound i; K_i: Freundlich parameter of compound i ($MM^{-1} M^{3/n} L^{-3/n}$).

Table 3. (Continued).

Adsorbent	Removal Target	ANN Architecture	Training Function	Layers No.	ANN Topology	Data No.	Input	Output	Epochs No.	Ref.
Porous polyethylene membrane	Bovine serum albumin	Feed-forward back propagation	–	3	2:6:1	36	Nondimensionalized depth and adsorption time	Nondimensionalized concentration	17280	[95]
Granular activated carbon	368 organic compounds	Feed-forward back propagation	–	3	4:3:1	368	Molecular size and flexibility, Molecular volume, Critical dimension, Dummy variable	*log(q_e/C_e)	–	[96]

* C_e: the solution concentration at equilibrium (mg/L); q_e: the adsorption capacity at equilibrium (mg/g).

Chapter 6

ANN MODELING OF BIOLOGICAL TREATMENT PROCESSES

Biological treatment of wastewaters is the use of bacteria and other micro/macro-organisms to remove contaminants by assimilating them. In fact, bacteria and other organisms use wastewater pollution as their nutrition and grow. Through their metabolism, the organic materials of wastewater are transformed into cellular mass, which is no longer in solution but can be precipitated at the bottom of a settling tank or retained as slime on solid surfaces or vegetation in the system. The water exiting the biological treatment system is then much clearer than it entered it [97]. In recent years a number of studies have been focused on some micro/macro-organisms that are able to biodegrade and biosorb contaminants in wastewaters. A wide variety of organisms are capable to treat a wide range of contaminants include bacteria, fungi, yeasts and algae [98-102]. The main advantages of biological treatment processes are:

- Operation at ambient temperature: there is no need to heat or cool the water, which saves on energy consumption. Because wastewater treatment operations take much space, they are located outdoor, and this implies that the system must be able to operate at seasonally varying temperatures;
- Low capital and operating costs compared to those of chemical-oxidation processes,
- True destruction of organics, versus mere phase separation, such as with air stripping or carbon adsorption;

- Oxidation of a wide variety of organic compounds;
- Removal of reduced inorganic compounds, such as sulfides and ammonia, and total nitrogen removal possible through denitrification;
- Operational flexibility to handle a wide range of flows and wastewater characteristics, and
- Reduction of aquatic toxicity.

Because of these advantages, many biological processes are used for treatment of wide variety of pollutants as reported in Table 4.

The major challenge in modeling of biological processes is the nonlinear and time varying nature of them. The modeling of nonlinear systems is not simple, and success has been limited to restrictive classes of nonlinear systems. In addition, the kinetic parameters of the various steps involved in the biological processes are very difficult to determine due to the complexity of the reactions. Thus, the application of the artificial neural networks to predict the performance of the biological systems has been attempted. ANN modeling of different biological water and wastewater treatment processes has been summarized in Table 4.

Khataee and co-workers have reported the ANN modeling of biological treatment of the dye solution using microaglae (*Chlamydomonas, Chlorella, Cosmarium* and *Euglena*) [102, 103] and macroalge (*Chara* and *Cladophora*) species [104, 105]. The three layer network with back propagation algorithm has been applied. The tan-sigmoid and linear transfer functions have been used as transfer functions in hidden and output layers, respectively. The samples have been divided into training, validation and test subsets that each of them contains 50, 25 and 25% of samples, respectively. The validation and test sets, for evaluation of the validation and modeling power of the model, have been randomly selected from the experimental data. Their findings indicated that ANN provides reasonable predictive performance ($R^2 > 0.95$). The relative importance of the various input variables on the output variables has been assessed on the basis of the neural net weight matrix. These results show that the developed ANN models can describe the behavior of the complex interaction biological process within the range of experimental conditions adopted.

Table 4. ANN modeling of biological water and wastewater treatment processes

Treatment Agent	Treatment Target	ANN Architecture	Training Function	Layers No.	ANN Topology	Data No.	Input	Output	Epochs No.	Ref.
Microalga *Chlamydomonas* sp.	Basic Green 4	Feed-forward back propagation	Train scaled conjugate gradient	3	5:16:1	435	$[Dye]_0$, $[Alga]_0$, Time, pH, Temperature	Biosorptive decolorization efficiency (%)	–	[102]
Three microalgae (*Chlorella*, *Cosmarium* and *Euglena* species)	Malachite Green	Feed-forward back propagation	Train scaled conjugate gradient	3	5:16:1	435	$[Dye]_0$, $[Alga]_0$, Time, pH, Temperature	Biological dye removal (%)	3000	[103]
Macroalga *Chara* sp.	Malachite Green	Feed-forward back propagation	Train scaled conjugate gradient	3	5:12:1	300	$[Dye]_0$, $[Alga]_0$, Time, pH, Temperature	Biological dye removal (%)	3000	[104]
Macroalga *Cladophora* sp.	Malachite Green	Feed-forward back propagation	Train scaled conjugate gradient	3	5:14:1	220	$[Dye]_0$, $[Alga]_0$, Time, pH, Temperature	Biological dye removal (%)	3000	[105]
Biosorption by sawdust	Cu(II)	Partial recurrent Back propagation	–	5	4:50:40:27:1	4864	$[Cu(II)]_0$, pH, Temperature, Particle size	Removal of Cu(II) ions (%)	1000	[106]
Microorganism *Pseudomonas* sp.	Phenol	Feed-forward and genetic algorithm	Real-coded genetic algorithm	3	1:5:1	–	–	Biodegradation of phenol	2000	[107]

Table 4. (Continued).

Treatment Agent	Treatment Target	ANN Architecture	Training Function	Layers No.	ANN Topology	Data No.	Input	Output	Epochs No.	Ref.
Bacteria *Ralstonia basilensis* and microalgae *Chlorella sorokiniana*	Salicylate	Feed-forward back propagation	Supervised hybrid algorithm	3	4:4:1	23	Light intensity, Temperature, Hydraulic retention time, [Salicylate]$_0$	Salicylate removal efficiency	–	[108]
Brown alga *Sargassum filipendula* sp.	Mixture of cadmium–zinc ions	–	–	3	7:5:2	–	Information about the equilibrium concentrations of each ion in the fluid phase	Cd(II) & Zn(II) adsorbed concentrations	–	[109]
Anaerobic sludge	46 kinds of aromatic compounds	Back propagation	–	3	5:5:1	41	Energy of the highest occupied molecular orbital, total energy, molar refractivity, the logarithm of the partition coefficient for n-octanol/water, standard Gibbs free energy	Integrated assessment indices	–	[110]
Modeling of Doha West wastewater treatment plant		Feed-forward	Levenberg–Marquardt	4	3:20:10:1	–	BOD, COD and TSS	BOD, COD or TSS	1000	[111]
				3	1:40:1					
Sulfidogenic treatment	Sulfate and zinc containing wastewater	Back propagation	Levenberg–Marquardt	3	5:20:4	160	Feed pH, sulfate, Zn^{2+}, COD and operation time	Effluent sulfate, COD, acetate and Zn^{2+} concentrations	–	[112]

Table 4. (Continued)

Treatment Agent	Treatment Target	ANN Architecture	Training Function	Layers No.	ANN Topology	Data No.	Input	Output	Epochs No.	Ref.
Microbial degradation	Biodegradability of mineral base oils	Feed-forward	Back propagation	3	6:3:1	31	Hydrocarbon classes of paraffins, naphthenes, aromatics, and sulfur aromatics, viscosity @40 °C and viscosity index	Biodegradability (%)	20000	[113]
	Dynamic modeling of yeast fermentation bioreactor	Feed-forward	Back propagation	3	8:14:1	–	Variations of the reactor temperature as a function of the cooling agent flow.	Bioreactor temperature versus time	50	[114]
	Biological bench-scale sequencing batch reactor	Feed-forward	Back propagation	4	6:10:5:3	–	* pH, ORP, DO, $[PO_4^{3-}]_0$, $[NO_3^-]_0$, $[NH_4^+]_0$	$[PO_4^{3-}]_t$, $[NO_3^-]_t$, $[NH_4^+]_t$	–	[115]
	Nitrogen and phosphorus									
	Control aeration in an aerated submerged biofilm wastewater treatment process	Adaptive neuro-fuzzy inference system: a multilayer feed-forward network which uses neural network learning algorithms and fuzzy reasoning to map inputs into an output.		5	3: 9: 27: 27:1	160	** COD_{in}, $NH_4^+{}_{in}$ and Q_{in}	** Rate of air supply (Q_{air}, m^3/h)	500	[66]

Table 4. (Continued).

Treatment Agent	Treatment Target	ANN Architecture	Training Function	Layers No.	ANN Topology	Data No.	Input	Output	Epochs No.	Ref.
Control aeration in an aerated submerged biofilm wastewater treatment process		Feed-forward	Back propagation	3	3:10:1	160	COD_{in}, $NH_4^+{}_{in}$ and Q_{in}	Rate of air supply (Q_{air}, m³/h)	500	[66]
Submerged membrane bioreactor	Cheese whey wastewater	Cascade-forward	Back propagation	3	8:3:4	30	Influent COD (mg/L), Solid retention time (day), Hydraulic retention time (day), Influent ammonia (mg/L), Influent nitrate (mg/L), Influent total phosphate (mg/L), Flux (L/m²/day), Pressure in membrane (bar)	Effluent COD (mg/L), Effluent ammonia (mg/L), Effluent nitrate (mg/L), Effluent total phosphate (mg/L)	1000	[116]

* **ORP**: Oxidation Reduction Potential; **DO**: Dissolved Oxygen.
** **Q_{air}**, defined as the airflow rate supplied by a blower for aeration in the biological treatment process (m³/h); **COD_{in}**: chemical oxygen demand (mg/L); **$NH_4^+{}_{in}$**: the ammonia concentration (mg/L); **Q_{in}**: the influent flow rate (defined as incoming flow rate to the treatment plant (m³/h)).

Yang et al. [110] have applied the stepwise regression and back propagation artificial neural network methods to establish quantitative structure biodegradability relationship (QSBR) of 46 kinds of aromatic compounds based on the assessment results. These aromatic compounds have been classified into readily, partially and poorly biodegradable compounds after calculation of their integrated assessment indices. In QSBR models, five molecular structure descriptors, energy of the highest occupied molecular orbital, total energy, molar refractivity, the logarithm of the partition coefficient for n-octanol/water, and standard Gibbs free energy, have been included. After analyzing the sensitivity of variables in QSBR models, it has been found that the key molecular structure descriptors affecting anaerobic biodegradability of aromatic compounds are total energy and molar refractivity, which are in direct proportion to the anaerobic biodegradability of aromatic compounds.

In another work, ANN has been applied for the prediction of biosorption efficiency of the removal of Cu(II) ions from aqueous solutions by sawdust. ANN model, based on multilayered partial recurrent back propagation algorithm, has been used to predict the extent of copper ions biosorption by taking into account the effect of initial Cu(II) concentration, pH, temperature and particle size of the adsorbent. The back propagation recurrent network using the momentum-training algorithm is found to be very effective to generalize and predict the degree of the adsorption. The configuration of the recurrent neural network that gives the best prediction is the one with three hidden layers consisting of 50, 40 and 27 neurons in each layer. ANN predicted results are very close to the experimental values. The average mean square error is 0.0021, which is sufficient to have error within ±1%. These results suggest that the proposed ANN model can be used as an effective technique in modeling, estimation and prediction of the biosorption process [106].

The biodegradation process of phenol in a fluidized bed bioreactor has been also simulated using genetic algorithm trained feed-forward neural network (See Table 4). Experiments have been carried out using the microorganism *Pseudomonas* sp. on synthetic wastewater. The steady state model equations describing the biodegradation process have been solved using feed-forward artificial neural network and genetic algorithm. The ANNs have been trained to a mean squared error level in the range of 10^{-4}. The minimization of the error function with respect to network parameters (weights and biases) has been considered as training of the network. Real-coded genetic algorithm has been used for training the network in an unsupervised manner.

The results suggest that the feed-forward neural network trained by real-coded genetic algorithm is a good technique for the simulation of the biodegradation process in a fluidized bed bioreactor [107].

Artificial neural networks are capable of inferring the complex relationships existing between input and output process variables without a detailed description of the mechanisms governing the process, and should therefore be more suitable for the modeling of photosynthetically oxygenated systems. For instance, a three layer neural network consisting of a single hidden layer with four neurons has been used to predict the steady-state operation of a continuous stirred tank photobioreactor during salicylate biodegradation by an algal-bacterial consortium. This designed network has exhibited a satisfactory fit for both training and testing data with correlation coefficients of 99%. This modeling approach is expected to contribute to improve the understanding of the complex relationships between light, temperature, hydraulic retention time, pollutant concentration and process removal efficiency, which would eventually promote the development of algal–bacterial processes as a cost effective alternative for the treatment of industrial wastewaters [108].

A reliable model for any real wastewater treatment plant is essential in order to provide a tool for predicting its performance and forming a basis for controlling the operation of the process. This would minimize the operation costs and assess the stability of environmental balance. This process is complex and attains a high degree of nonlinearity due to the presence of bio-organic constituents that are difficult to model using mechanistic approaches. Predicting the plant operational parameters using conventional experimental techniques is also a time consuming step and is an obstacle in the way of efficient control of such processes. Two ANN models with one or two hidden layers have been used to the modeling of a real wastewater plant in Doha [111]. The ANN model provided accurate predictions of the effluent stream, in terms of biological oxygen demand (BOD), chemical oxygen demand (COD) and total suspended solids (TSS) when using COD as an input in the crude supply stream.

In another work, a submerged membrane bioreactor receiving cheese whey has been modeled by artificial neural network and its performance over a period of 100 days at different solids retention times has been evaluated. A cascade-forward network has been used to model the membrane bioreactor and normalization has been used as a preprocessing method. The network has been fed with two subsets of operational data, with two-thirds being used for training and one-third for testing the performance of the artificial neural network. The modeling procedure for effluent chemical oxygen demand, ammonia, nitrate and total phosphate concentrations has been very successful and a perfect match has been obtained between the measured and the calculated concentrations [116].

From the above studies, it can be concluded that artificial neural networks provide a robust tool for predicting the performance of biological wastewater treatment processes even though such systems involve highly complex physical and biochemical mechanisms controlled by the micro/macro-organisms. Furthermore, ANN models can do this without the requirement for conversion of operational data or measurement of additional operational parameters.

Chapter 7

ANN MODELING OF ELECTROCHEMICAL TREATMENT PROCESSES

Electrochemical technologies can be applied for the treatment of effluents released from a wide range of industries or processes. These techniques have been receiving greater attention in recent years due to their distinctive advantages such as environmental compatibility (the main reactant is the electron which is a clean reagent), and versatility (a plethora of reactors and electrode materials, shapes, and configurations can be utilized). It is noteworthy that the same reactor can be used frequently for different electrochemical reactions with only minor changes and also the electrolytic processes can be scaled easily from the laboratory to the plant, allowing treatment volumes ranging from milliliters to millions of liters. Electrochemical methods are generally safe because of the mild conditions and innocuous nature of the chemicals usually employed. Electrodes and cells can be designed to minimize power losses due to poor current distribution and voltage drops and making the processes more competitive in energy consumption than the conventional techniques [117].

Artificial neural networks have been widely used for modeling of electrochemical methods of water and wastewater treatment (see Table 5).

For example, Basha et al. [118] have applied ANN method for modeling the electrochemical degradation in batch, batch recirculation and continuous recycle modes in managing the wastewater discharged by a typical medium-scale specialty chemical industry. They have developed three and four layers back propagation feed-forward nets trained with Levenberg-Marquardt algorithm. It has been reported that single hidden layer feed-forward back propagation neural network is adequate enough to predict the performance of the process. Daneshvar et al. [119] have used a three layer feed-forward back propagation neural network for modeling the electrocoagulation removal of C.I. Basic Yellow 28.

Electrocoagulation (EC) as an electrochemical method has been developed to overcome the drawbacks of conventional water and wastewater treatment technologies. EC process provides a simple, reliable and cost-effective method for the treatment of wastewater without any need for additional chemicals, and thus the secondary pollution. It also reduces the amount of sludge, which needs to be disposed [120-122].

Daneshvar and co-workers have reported that ANN modeling could be successfully used to investigate the cause effect relationship in electrocoagulation process [119]. The ANN model could describe the behavior of the complex reaction system with the range of experimental conditions adopted. Simulation based on the ANN model can estimate the behavior of the system under different conditions.

Aber et al. [123] have also developed a three layer feed-forward back propagation neural network for modeling of the electrocoagulation removal of Cr(VI) from the polluted solutions. The network was developed with sigmoidal transfer function as a transfer function in the hidden and output layers. 212 experimental sets have been used to develop the ANN model. The linear regression between the network prediction and the corresponding experimental data (R^2=0.976) proves that modeling of the electrocoagulation removal of Cr(VI) using artificial neural network is a good and precise method to predict the residual concentration of Cr(VI) under different conditions.

A five layer (4:10:10:10:1) back propagation feed-forward network has been also developed to predict the electrocoagulation removal of Acid Blue 113. The training function was *trainlm* and totally 144 experimental sets have been used to develop the ANN model. The authors have reported that ANN model predictions satisfactorily match with experimental observation [124].

One of the most popular electrochemical advanced oxidation process (EAOP) is anodic oxidation (AO) consisting in the destruction of organics in an electrolytic cell under the action of hydroxyl radical formed as intermediate from water oxidation to O_2 at the surface of a high O_2-overvoltage anode:

$$M + H_2O \rightarrow M(OH^\bullet) + H^+ + e^- \qquad (23)$$

where $M(OH^\bullet)$ denotes the hydroxyl radical adsorbed on the anode M or remaining near its surface [125].

More potent indirect electro-oxidation methods with hydrogen peroxide electrogeneration are being also developed for wastewater remediation. In these techniques, H_2O_2 is continuously supplied to the contaminated solution from the two-electron reduction of O_2 usually at carbon-felt [126-131] and carbon-polytetrafluoroethylene O_2-diffusion [132, 133] cathodes:

$$O_{2(g)} + 2H^+ + 2e^- \rightarrow H_2O_2 \tag{24}$$

The electro-Fenton process performs when Fe^{2+} is added to the solution.

$$Fe^{2+} + H_2O_2 + H^+ \rightarrow Fe^{3+} + OH^\bullet + H_2O \tag{25}$$

The peroxi-coagulation process is performed with a sacrificial Fe anode, which continuously supplies soluble Fe^{2+} to the solution from the following anodic oxidation reaction [134]:

$$Fe \rightarrow Fe^{2+} + 2e^- \tag{26}$$

Fe^{2+} thus produced is quickly oxidized by electrogenerated H_2O_2 from reaction (24) yielding a solution saturated with Fe^{3+}, whereas the excess of this ion precipitates as hydrated Fe(III) oxide (Fe(OH)$_3$).

$$Fe^{2+} + H_2O_2 \rightarrow Fe(OH)^{2+} + OH^\bullet \tag{27}$$

Then, pollutants can be removed by the combined action of their degradation with $OH^\bullet$ generated from reaction (25) and their coagulation with the Fe(OH)$_3$ precipitate formed.

Table 5. ANN modeling of electrochemical water and wastewater treatment processes

Treatment Process	Treatment Target	ANN Architecture	Training Function	Layers No.	ANN Topology	Data No.	Input	Output	Epochs No.	Ref.
Peroxi-coagulation	C.I. Basic blue 3, Malachite green, C.I. Basic red 46, C.I. Basic yellow 2	Feed-forward back propagation	trainscg	3	4:14:1	60	Electrolysis time, Initial pH, Applied current, $[Dye]_0$	Decolorization (%)	–	[9]
Peroxi-coagulation	C.I. Basic yellow 2	Feed-forward back propagation	trainscg	3	4:16:1	117	Electrolysis Time, Initial pH, Applied current, $[Dye]_0$	Decolorization (%)	–	[22]
Electrochemical oxidation	Real wastewater (Wastewater of Specialty Chemical Industry)	Feed-forward back propagation	Levenberg–Marquart	3	3:5:1 3:7:1 3:9:1	124	Current density, Electrolysis duration, Supporting electrolyte concentration	COD removal (%)	–	[118]
				4	3:3:3:1 3:3:5:1 3:3:7:1					
Electro-coagulation	C. I. Basic yellow 28	Feed-forward back propagation	trainscg	3	7:10:1	49	Current density, Electrolysis Time, Initial pH, $[Dye]_0$, Conductivity, Retention time, Inter-electrodes distance	Decolorization (%)	–	[119]

Treatment Process	Treatment Target	ANN Architecture	Training Function	Layers No.	ANN Topology	Data No.	Input	Output	Epochs No.	Ref.
Electro-coagulation	Cr(IV)	Feed-forward back propagation	–	3	4:10:1	212	Current density, Time of electrolysis, $[Cr(VI)]_0$, Concentration of electrolyte	$[Cr(VI)]_t$	–	[123]
Electro-coagulation	Acid Blue 113	Feed-forward back propagation	trainlm	5	4:10:10:10:1	144	Effluent concentration, Electrolyte pH, Current density, Electrolysis time	COD removal (%)	–	[124]
Electrolysis	Phenolic compounds (phenol, 4-chlorophenol, 2,4-dichlorophenol, 2,4,6-trichlorophenol, 4-nitrophenol and 2,4-dinitrophenol)	Feed-forward back propagation	Different functions	4	7:30:25:1 7:25:20:1	420	Temperature, $[COD]_0$, pH, Current density, Current charge passed, Types of chlorine phenol compound, Type of nitrophenols compounds	COD removal (%)	–	[136]
Electrodialysis	Pb^{2+}	Multilayer perceptron	Levenberg–Marquardt	4	4:5:4:1	81	$[Pb^{2+}]_0$ Temperature, Flow rate, Voltage	$[Pb^{2+}]$	–	[137]
Coagulation	Drinking water	Self organizing map	–	4	7:8:0:3	202	Turbidity, Color, Absorption 254, Residual aluminum, pH, Alum dose, DOC	Turbidity, Color, Absorption 254 nm, Residual aluminum	–	[138]

Peroxi-coagulation differs from classical electrocoagulation with a Fe anode and a graphite cathode, since electrocoagulation does not degrade soluble organics because no significant H_2O_2 is produced in the medium, such as previously found for aniline and 4-chlorophenol [134, 135].

Salari et al. [22] have developed a feed-forward back propagation neural network for modeling of peroxi-coagulation removal of C.I. Basic Yellow 2. The network is developed with sigmoidal transfer function as a transfer function in the hidden and output layers. The three layer network was trained with scaled conjugate gradient algorithm using totally 117 data sets. The linear regression between the network prediction and the corresponding experimental data (R^2=0.9713) proves that modeling the peroxi-coagulation removal of BY2 using artificial neuron network is a satisfactory method. The authors have concluded that artificial neural network modeling has been successfully used to investigate the cause effect relationship in peroxi-coagulation process.

In a similar work, Zarei et al. [9] have used neural network modeling to predict the performance of a peroxi-coagulation removal of four dyes namely Basic Blue 3, Malachite Green, Basic Red 46 and Basic Yellow 2 using carbon nanotube-PTFE cathode. They have used 60 data sets to develop a three layer (4:14:1) feed-forward network to predict the decolorization efficiency of the process as the output of the neural network. The linear regression between the network prediction and the corresponding experimental data (R^2=0.989) show that ANN modeling the peroxi-coagulation removal of Basic Yellow 2 as a model dye is a precious method.

Neural networks have also been developed to model the electrolysis of wastes polluted with phenolic compounds, including phenol, 4-chlorophenol, 2,4-dichlorophenol, 2,4,6-trichlorophenol, 4-nitrophenol and 2,4-dinitrophenol [136]. In this work, Piuleac and co-workers have proposed feed-forward four layers networks (7:30:25:1 and 7:25:20:1) using different training algorithms to predict COD content of the treated solutions.

Sadrzadeh et al. [137] have developed a neural network to predict the separation of lead ions from wastewater using electrodialysis method. In this work, 81 data sets were used to train a multilayer perceptron network using Levenberg-Marquardt algorithm. The developed four layers network (4:5:4:1) can effectively predict lead ions concentration in the treated solution. The authors have reported that ANN modeling technique have many favorable features such as efficiency, generalization and simplicity, which make it an attractive choice for modeling of complex systems, such as wastewater treatment processes.

It can be concluded that artificial neural networks are capable to simulate the complex relationships existing between input and output process variables in electrochemical methods. ANNs can overcome the difficulty of modeling such processes where different phenomena such as mass and heat transfers and mechanic fluids are involved to run the overall process.

Chapter 8

ANN MODELING OF PHOTOCATALYTIC PROCESSES

In recent years, "advanced oxidation processes" (AOPs) have been developed as an alternative to the conventional water and wastewater treatment methods. These processes are based on the generation of very reactive species such as hydroxyl radicals that have been proposed to oxidize quickly and nonselectively a broad range of organic pollutants [139–143]. Different AOPs have been schematically indicated in Figure 12. AOPs have been developed to degrade the nonbiodegradable contaminants of water into harmless species (e.g. CO_2, H_2O, etc).

Heterogeneous photocatalysis via combination of a photocatalyst (e.g. TiO_2 and ZnO) and UV light is considered one of the promising advanced oxidation processes for destruction of soluble organic pollutants found in water and wastewater [141, 143]. Some of the beneficial characteristics of TiO_2 in comparison to other photocatalysts include high photocatalytic efficiency, physical and chemical stability, low cost and low toxicity [144]. Interaction of TiO_2 with photons that possess energy equal or higher to the band gap may cause separation of conduction and valence bands as illustrated in Figure 13. This event is known as electron–hole pair generation. For TiO_2, this energy can be supplied by photons with energy in the near ultraviolet range. This property promotes TiO_2 as a promising candidate in photocatalysis where solar light can be used as the energy source. Indeed, when TiO_2 is illuminated with $\lambda<390$ nm light, an electron excites out of its energy level and consequently leaves a hole in the valence band. As electrons are promoted from the valence band to the conduction band, they generate electron–hole pairs (Eq. (28)) [144–146]:

$$TiO_2 + hv(\lambda < 390nm) \rightarrow e^- + h^+ \tag{28}$$

Valence band (h^+) potential is positive enough to generate hydroxyl radicals ($OH^\bullet$) at TiO_2 surface and the conduction band (e^-) potential is negative enough to reduce molecular oxygen as shown in the Figure 13.

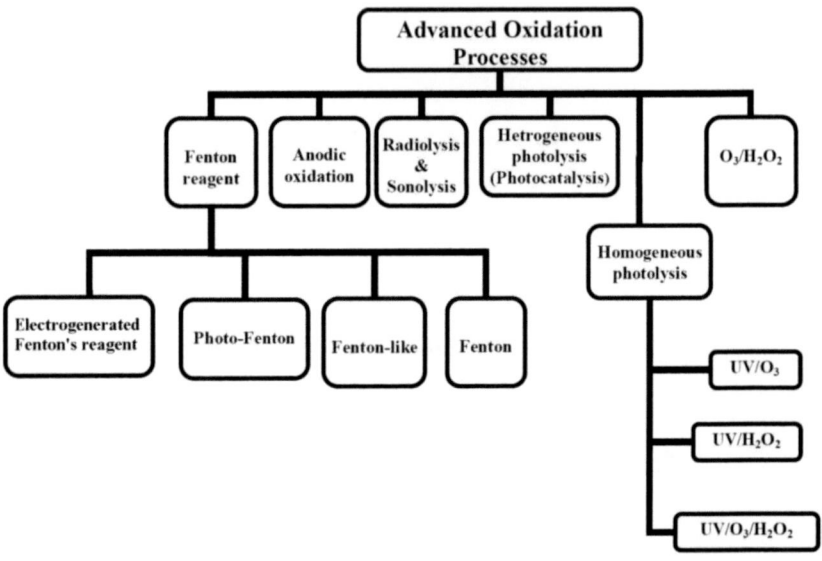

Figure 12. Advanced Oxidation Processes (AOPs).

The hydroxyl radical is a powerful oxidizing agent which may attack the organic matters present at or near the surface of TiO_2. It is capable to degrade toxic and bioresistant compounds into harmless species (e.g. CO_2, H_2O, etc). TiO_2 nanomaterials are successfully used for the photocatalytic remediation of a variety of organic pollutants such as hydrocarbons and chlorinated hydrocarbons (e.g. CCl_4, $CHCl_3$, C_2HCl_3, phenols, chlorinated phenols, surfactants, pesticides, and dyes) as well as reduction deposition of heavy metals such as Pt^{4+}, Pd^{2+}, Au^{3+}, Rh^{3+} and Cr^{3+} from aqueous solutions. TiO_2 nanomaterials have also been effective in the destruction of biological organisms such as bacteria, viruses, and molds [147–151].

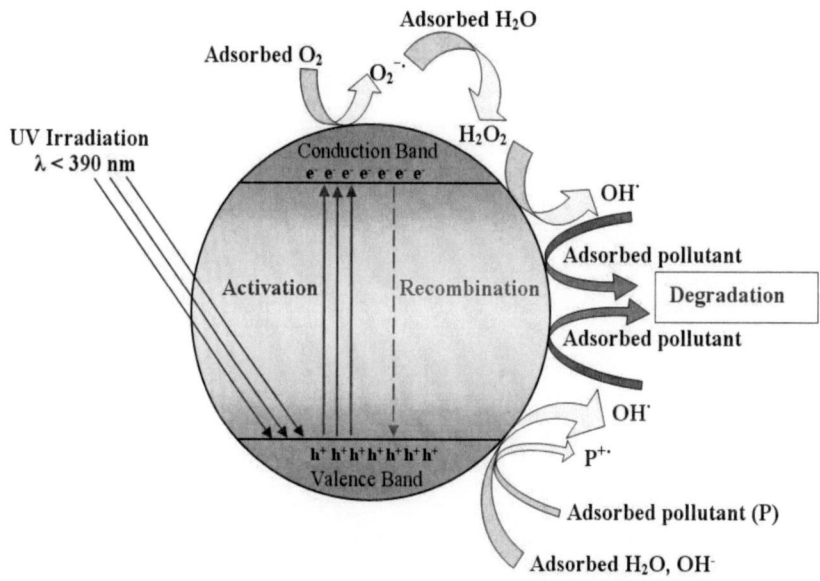

Figure 13. Generation of photocatalytic active species at the surface of TiO_2 nanoparticles.

Due to the complexity of the reactions in the photocatalytic processes, the kinetic parameters of the various steps involved are very difficult to determine, leading to uncertainties in the design and scale-up of chemical reactors of industrial interest. This is caused by the complexity of solving the equations that involve the radiant energy balance, the spatial distribution of the absorbed radiation, mass transfer, and mechanisms of a photocatalytic degradation process involving radical species. Because of these reasons, the modeling of the photocatalytic processed via artificial neural networks is quite appropriate. As it was mentioned, one of the characteristics of modeling based on ANNs is that it does not require the mathematical description of the phenomena involved in the process. The examples of ANN modeling of photocatalytic water and wastewater treatment processes are summarized in Table 6.

Table 6. ANN modeling of photocatalytic water and wastewater treatment processes

Treatment Process	Treatment Target	ANN Architecture	Training Function	Layers No.	ANN Topology	Data No.	Input	Output	Epochs No.	Ref.
UV/TiO_2	Nitrilotriacetic acid (NTA)	Feed-forward back propagation	–	3	3:5:1	58	$[TiO_2]_0$, $[NTA]_0$, Time	NTA degradation (%)	5000	[71]
Solar/Fenton/TiO_2	Imipramine	Feed-forward back propagation	Conjugate gradient descent	3	3:3:1	16	$[H_2O_2]_0$, $[Fe(II)]_0$, $[TiO_2]_0$	Imipramine degradation (%)	95	[152]
UV/TiO_2	Nitrogen oxides (NO and NO_x)	Feed-forward	Quick propagation	5	3:7:4:3:2	488	$[TiO_2]_0$, Exposed surface, Time	NO and NO_x degradation (%)	2000	[153]
Solar/Fenton/TiO_2	Reactive blue 4	Feed-forward back propagation	Marquardt non-linear	–	–	19	pH, $[TiO_2]_0$, $[Fe(II)]_0$, $[H_2O_2]_0$	Decolorization kinetic constant (min^{-1})	–	[154]
Solar/H_2O_2/TiO_2	IGCC power station effluents	Feed-forward back propagation	Marquardt non-linear	–	–	11	$[TiO_2]_0$, $[H_2O_2]_0$	Cyanide and Formates degradation constant (min^{-1})	–	[155]
Solar/H_2O_2/ZnO	IGCC power station effluents	Feed-forward back propagation	Marquardt non-linear	–	–	11	$[ZnO]_0$, $[H_2O_2]_0$	Cyanide and Formates degradation constant (min^{-1})	–	[156]

Table 6. (Continued)

Treatment Process	Treatment Target	ANN Architecture	Training Function	Layers No.	ANN Topology	Data No.	Input	Output	Epochs No.	Ref.
UV/Immobilized–TiO_2	C. I. Basic red 46	Feed-forward back propagation	Train scaled conjugate gradient	3	4:8:1	140	pH, $[Dye]_0$, Time, UV light intensity	Decolorization (%)	5000	[157]
UV/TiO_2	2,4-dihydroxybenzoic acid (DHBA)	Feed-forward back propagation	–	3	3:4:1	66	$[DHBA]_0$, $[TiO_2]_0$, Time	$[DHBA]_t$ (mg/L)	20000	[158]
UV/TiO_2	Ethylene diamine tetraacetic acid (EDTA)	Feed-forward back propagation	–	3	3:4:4	26	$[TiO_2]_0$, $[EDTA]_0$, pH	EDTA degradation at 30, 60, 90, and 120 min	50	[159]

Artificial neural networks have been used for modeling of TiO_2 photocatalytic degradation of 2,4-dihydroxybenzoic acid, chosen as a model water contaminant, as a function of the concentrations of substrate and catalyst. The experimental design methodology has been applied to the choice of an appropriate set of experiments well distributed in the experimental region (Doehlert uniform array). Contrary to a classical treatment of the data, based on apparent rate constants modeled by a quadratic polynomial function, neural network analysis of the same experimental data does not require the use of any kinetic or phenomenological equations and allows the simulation and the prediction of the pollutant degradation as a function of irradiation time, as well as prediction of reaction rates, under varying conditions within the experimental region [158].

An artificial neural network model has been developed to predict the photocatalytic decolorization of C.I. Basic Red 46 solution. This dye, commonly used as a textile dye, has been photocatalytically removed using supported TiO_2 nanoparticles irradiated by a 30 W UV-C lamp in a batch reactor. The photocatalyst has been industrial Degussa P25 (crystallite mean size 21 nm) immobilized on glass beads by a heat attachment method. The process of the dye decolorization in the presence of TiO_2 nanoparticles has been experimentally studied through changing the initial dye concentration, UV light intensity and initial pH. The influence of inorganic anions such as chloride, sulfate, bicarbonate, carbonate and phosphate on the photocatalytic decolorization of the dye has been investigated. The findings indicated that the proposed ANN model provides reasonable predictive performance (R^2=0.96). The influence of each parameter on the variable studied has been assessed: initial concentration of the dye being the most significant factor, followed by the initial pH and reaction time [157].

A chemometric study on the TiO_2 photocatalytic degradation of nitrilotriacetic acid (NTA) in aqueous media under UV radiation has been carried out taking into account the multiple variables that take part in the system. To save redundant number of experiments, the system has been managed under chemometric techniques for several variables as NTA and TiO_2 concentrations, pH and irradiation time. Multiple-way analysis of the variance (MANOVA) has been applied to find the statistically significant variables. An artificial neural network has been used to build an empirical model of the system. All measurements have been driven under experimental designs: a full-factorial design has been used to analyze significant factors through MANOVA, and a Doehlert design, which was modified by spatial rotation, was applied in order to have a satisfactory number of levels for the factor time to be able to train the ANN. The study allows the knowledge and prediction of the behavior of the system as well as to work out kinetic parameters and to optimize their variables. The results of kinetic parameters obtained with the neural network agree with independent experimental results [71].

Toma and co-workers [153] have been proposed an artificial neural network model to analyze the photocatalytic removal of nitrogen oxides over TiO_2 powders. It should be mentioned that the nitrogen oxides NO_x (especially NO and NO_2) forming by cars traffic, combustion of coals and thermal power plants participate in the formation of acid rain, greenhouse effect, photochemical pollution and major problems on the human health. 488 experimental sets have been used to feed the ANN model. The network input contains three neurons representing TiO_2 powder quantity (g), reaction time (min) and exposed surface (cm^2). The output pattern comprise two neurons representing the photocatalytic response, namely NO and NO_x efficiencies. Experimental sets are organized in training and test samples. The first category is used to tune neuron network weights and the second category to test the network configuration. A stopping criterion is applied corresponding to a fixed number of training and test cycles: 2000 cycles are achieved for each network configuration after approximately 1 h of computation time. The structure of the optimized ANN model characterize by three hidden layers containing seven, four and three neurons, respectively (See Table 6). Predicted results are in good agreement with experimental ones (R^2=0.9857). With the ANN optimized structure, it is possible to quantify the effect of each experiment variable by varying independently each of them and collecting NO and NO_x efficiencies [153].

Emilio and co-workers [159] have been also used the chemometric techniques including full factorial and Doehlert experimental designs, multivariate analysis by MANOVA and artificial neural networks for the photocatalytic reaction of ethylenediaminetetraacetic acid (EDTA) over TiO_2 in aqueous solution. EDTA concentration, TiO_2 amount, pH of the solution and irradiation time have been chosen to build a set of experiments for the analysis. Correlation plots among variables have been built a model for prediction the behavior of the photocatalytic system and optimizing parameters.

The heterogeneous assisted photocatalytic degradation processes (i.e. Solar/H_2O_2/TiO_2 and Solar/H_2O_2/ZnO) of wastewater from a thermoelectric power station under concentrated solar light irradiation using a Fresnel lens has been also reported [155, 156]. The efficiency of photocatalytic degradation processes has been determined from the analysis of cyanide and formate removal. The experimental kinetic constants have been fitted using neural networks. The ANN applied has been solved with two neurons and using a simple exponential activation function and the strategy is based on a back propagation calculation. Input variables have been initial concentration of hydrogen peroxide and photocatalyst while output data have been cyanides and formates degradation constant. The analysis of the relevance of each variable with respect to the others has been reported. It has been found that the initial concentration of hydrogen peroxide is the most significant factor affecting the degradation kinetic rate constants of cyanide and formate [155].

From the above studies, it can be concluded that the artificial neural networks can describe the behavior of the complex reaction system such as photocatalytic processes in the range of experimental conditions adopted. Simulation based on the ANN models can estimate the behavior of the photocatalytic processes under different conditions.

Chapter 9

ANN MODELING OF PHOTOOXIDATIVE PROCESSES

As it was mentioned, advanced oxidation processes are the alternative to the conventional water and wastewater treatment methods. Hydroxyl radicals ($^{\bullet}$OH), highly reactive species generated in sufficient quantities by these systems, have the ability to oxidize the majority of the organics in the polluted waters [160]. As shown in chapter 8, common AOPs involve Fenton process, anodic oxidation, radiolysis and sonolysis, ozonation, heterogeneous and homogeneous photolysis with H_2O_2 and O_3, and various combinations of these methods [161, 162].

Fenton and photo-Fenton processes have been widely used for destruction of organics in the polluted waters. Besides, UV/H_2O_2, ozonation and chlorination are also some of the most important AOPs used for treatment of water and wastewater. In the following sections, a brief mechanistic description of these techniques as well as the results of ANN modeling will be presented.

9.1. FENTON AND PHOTO–FENTON PROCESSES

In 1876, Fenton's pioneering work pointed out the possible use of a mixture of H_2O_2 and Fe^{2+} to destroy tartaric acid. Most people, however, consider that Fenton's chemistry began in 1894 when he published a deeper study on the strong promotion of the oxidation of this acid with such a reagent. During the period 1901-1928, the stoichiometry of the reaction between H_2O_2

and Fe^{2+} has been studied. The extraordinary practical usefulness of Fenton's reagent for the oxidation of organic compounds was first assumed in the 1930s as a radical mechanism for the catalytic decomposition of H_2O_2 by iron salts [163].

Hydrogen peroxide (H_2O_2) is a strong oxidant (standard potential 1.80 and 0.87 V at pH 0 and 14, respectively) [164] and its application in the treatment of various inorganic and organic pollutants is well established. Numerous applications of H_2O_2 in the removal of pollutants from wastewater, such as sulfites, hypochlorites, nitrites, cyanides, and chlorine have been reported [165].

Oxidation by H_2O_2 alone is not effective for high concentrations of certain refractory contaminants, such as highly chlorinated aromatic compounds and inorganic compounds, because of low rates of reaction at reasonable H_2O_2 concentrations. Transition metal salts (e.g. iron salts), ozone and UV–light can activate H_2O_2 to form hydroxyl radicals (Eqs. (29-31)) which are strong oxidants:

- Ozone and hydrogen peroxide

$$O_3 + H_2O_2 \rightarrow OH^{\bullet} + O_2 + HO_2 \tag{29}$$

- Iron salts and hydrogen peroxide

$$Fe^{2+} + H_2O_2 \rightarrow Fe^{3+} + OH^{\bullet} + OH^{-} \quad k_1 \approx 70 M^{-1} s^{-1} \text{ [166]} \tag{30}$$

- UV-light and hydrogen and hydrogen peroxide

$$H_2O_2 + UVC \rightarrow 2OH^{\bullet} \tag{31}$$

Fenton's reagent was discovered about 100 years ago, but its application as an oxidizing process for destroying toxic organics was not applied until the late 1960s [167]. Fenton reaction wastewater treatment processes are known to be very effective in the removal of many hazardous organic pollutants from water [163]. The main advantage is the complete destruction of contaminants to harmless compounds, such as CO_2, water and inorganic salts. The Fenton reaction causes the dissociation of the oxidant and the formation of highly reactive hydroxyl radicals that attack and destroy the organic pollutants [163].

Fenton's reagent is a mixture of H_2O_2 and ferrous iron (Fe^{2+}), which generates hydroxyl radicals according to reaction (30) [168-170]. The ferrous iron initiates and catalyzes the decomposition of H_2O_2, resulting in the generation of hydroxyl radicals (chain initiation). The generation of the radicals involves a complex reaction sequence in an aqueous solution (Eqs. (30, 32)) [171].

$$OH^\bullet + Fe^{2+} \rightarrow OH^- + Fe^{3+} \text{ (Chain termination)} \quad (32)$$
$$k_2 = 3.2 \times 10^8 \, M^{-1} s^{-1}$$

Moreover, the newly formed ferric ions may catalyze hydrogen peroxide, causing it to be decomposed into water and oxygen. Ferrous ions and radicals are also formed in the reactions (see reactions (33-37)) [172].

$$Fe^{3+} + H_2O_2 \leftrightarrow Fe-OOH^{2+} + H^+$$
$$k_3 = 0.001 - 0.01 \, M^{-1} s^{-1} \quad (33)$$

$$Fe-OOH^{2+} \rightarrow HO_2^\bullet + Fe^{2+} \quad (34)$$

The reaction of hydrogen peroxide with ferric ions is referred to as a Fenton–like reaction (Eqs. (31, 34)) [171-174].

$$Fe^{2+} + HO_2^\bullet \rightarrow Fe^{3+} + HO_2^- \quad (35)$$
$$k_4 = 1.3 \times 10^6 \, M^{-1} s^{-1} \text{ at } pH = 3$$

$$Fe^{3+} + HO_2^\bullet \rightarrow Fe^{2+} + O_2 + H^+ \quad (36)$$
$$k_5 = 1.2 \times 10^6 \, M^{-1} s^{-1} \text{ at } pH = 3$$

$$OH^\bullet + H_2O_2 \rightarrow H_2O + HO_2^\bullet \quad (37)$$
$$k_6 = 3.3 \times 10^7 \, M^{-1} s^{-1}$$

As seen in reaction (37), H_2O_2 can act as an $OH^\bullet$ scavenger as well as an initiator. Hydroxyl radicals can oxidize organics (RH) by abstraction of protons producing organic radicals ($R^\bullet$), which are highly reactive and can be further oxidized (Eq. (38)) [175, 176].

$$RH + OH^\bullet \rightarrow H_2O + R^\bullet \rightarrow \text{Further oxidation} \qquad (38)$$

Since $k_6 = 10^7$ M^{-1} s^{-1} while $k_2 > 10^8$, reaction (37) can be made unimportant by maintaining a high $[RH]/[H_2O_2]$ ratio.

If the concentrations of reactants are not limiting, the organics can be completely detoxified by full conversion to CO_2, water and in the case of substituted organics, inorganic salts if the treatment is continued.

Walling [177] simplified the overall Fenton chemistry (Eq. (30)) by accounting for the dissociation water:

$$2Fe^{2+} + H_2O_2 + 2H^+ \rightarrow 2Fe^{3+} + 2H_2O \qquad (39)$$

The reaction (39) suggests that the presence of H^+ is required in the decomposition of H_2O_2, indicating the need for an acid environment to produce the maximum amount of hydroxyl radicals. Previous Fenton studies have shown that acidic pH levels near 3 are usually optimum for Fenton oxidations [178]. In the presence of organic substrates, excess ferrous ion, and at low pH, hydroxyl radicals can add to the aromatic or heterocyclic rings (as well as to the unsaturated bonds of alkenes or alkynes).

Hydroxyl radicals can also abstract a hydrogen atom, initiating a radical chain oxidation (Eqs. (40-42)) [177, 179].

$$RH + OH^\bullet \rightarrow H_2O + R^\bullet \quad \text{(Chain propagation)} \qquad (40)$$

$$R^\bullet + H_2O_2 \rightarrow ROH + OH^\bullet \qquad (41)$$

$$R^\bullet + O_2 \rightarrow ROO^\bullet \qquad (42)$$

The organic free radicals ($R^\bullet$) produced in reaction (40) may then be oxidized by Fe^{3+}, reduced by Fe^{2+}, or dimerized according to the following reactions (Eqs. (43-45)) [180].

$$R^\bullet + Fe^{3+} \xrightarrow{-oxidation} R^+ + Fe^{2+} \qquad (43)$$

$$R^\bullet + Fe^{3+} \xrightarrow{-reduction} R^- + Fe^{2+} \quad (44)$$

$$2R^\bullet \xrightarrow{-dimerization} R-R \quad (45)$$

The sequence of reactions (30), (32), (40) and (43) constitute the present accepted scheme for the Fenton's reagent chain. The ferrous ions generated in the above redox reactions (43) and (44) react with hydroxide ions to form ferric hydroxo complexes according to the following reactions (Eqs. (46-50)) [175, 176].

$$[Fe(H_2O)_6]^{3+} + H_2O \leftrightarrow [Fe(H_2O)_5 OH]^{2+} + H_3O^+ \quad (46)$$

$$[Fe(H_2O)_5 OH]^{2+} + H_2O \leftrightarrow [Fe(H_2O)_4 (OH)_2]^+ + H_3O^+ \quad (47)$$

Within pH 3 and 7, the above complexes become:

$$2[Fe(H_2O)_5 OH]^{2+} \leftrightarrow [Fe(H_2O)_8 (OH)_2]^{4+} + 2H_2O \quad (48)$$

$$[Fe(H_2O)_8 (OH)_2]^{4+} + H_2O \leftrightarrow [Fe_2(H_2O)_7 (OH)_3]^{3+} + H_3O^+ \quad (49)$$

$$[Fe_2(H_2O)_7 (OH)_3]^{3+} + [Fe(H_2O)_5 OH]^{2+} \leftrightarrow [Fe_2(H_2O)_7 (OH)_4]^{5+} + 2H_2O \quad (50)$$

which accounts for the coagulation capability of Fenton's reagent. Dissolved suspended solids are captured and precipitated. It should be noted that large amounts of small flocs are consistently observed in the Fenton oxidation step. Those flocs take a very long time, sometimes overnight, to settle out. Chemical coagulation using polymer is therefore necessary.

Fenton's reagent is known to have different treatment functions, as mentioned earlier, depending on the $H_2O_2/FeSO_4$ ratio. When the amount of Fe^{2+} employed exceeds that of H_2O_2, the treatment tends to have the effect of chemical coagulation. When the two amounts are reversed, the treatment tends to have the effect of chemical oxidation.

Reaction (43) competes with both the chain termination reaction (Eq. (32)) and with the propagation reaction (Eq. (39)) of Fenton chemistry. This competition for hydroxyl radical between Fe^{2+}, RH and Fe^{3+} leads to the non–

productive decomposition of hydrogen peroxide and limits the yield of hydroxylated (oxidized) organic compounds. Therefore, the stoichiometric relationship between Fe^{2+}, RH and Fe^{3+} has to be established to maximize the efficiency of the degradation process.

The wastewater treatment applying Fenton and photo–Fenton processes is, in general, quite complex. This is caused by the complexity of solving the equations that involve the radiant energy balance, the spatial distribution of the absorbed radiation, mass transfer, and the mechanisms of a photochemical or photocatalytic degradation involving radical species. Since the process depends on several factors, modeling of these processes involves many problems. It is evident that these problems cannot be solved by simple linear multivariate correlation. Artificial neural networks are now commonly used in many areas of chemistry and they represent a set of methods that may be useful in solving such problems (See Table 7) [7, 8, 181-183].

Elmolla et al. [184] have used neural network modeling to predict the performance of Fenton process for removal of antibiotics (amoxicillin, ampicillin and cloxacillin). The configuration of the back propagation neural network giving the smallest MSE was three layers ANN with tangent sigmoid transfer function (tansig) at hidden layer with 14 neurons, linear transfer function (purelin) at output layer and Levenberg–Marquardt back propagation training algorithm (LMA). ANN predicted results were very close to the experimental results with correlation coefficient (R^2) of 0.997 and MSE 0.000376.

Yu et al. [185] have also developed a novel Fenton process control strategy using ANN models and oxygen-reduction potential (ORP) monitoring to treat two synthetic textile wastewaters containing two common dyes namely Reactive Blue 49 (RB49) and Reactive Brilliant Blue (RBB). Experimental results indicate that the ANN models can predict precisely the color and chemical oxygen demand (COD) removal efficiencies for synthetic textile wastewaters with correlation coefficients of 0.91–0.99.

When UV irradiation is combined with some powerful oxidant, such as H_2O_2, organic dye degradation efficiency can be significantly enhanced due to hydroxyl radical generation caused by the photolysis of H_2O_2 (Eq. (31)), and these highly reactive non–selective radicals may further react with the organic substrate [186]. This process demonstrated high efficiency in the treatment of different types of organic dyes [187–190]. The efficiency of Fenton process could be significantly increased under light irradiation, where Fe^{3+} ions are constantly reduced to the Fe^{2+}, (Eq. (51)), [191] and the Fenton process is improved by the participation of photogenerated Fe^{2+}:

$$Fe^{3+} + H_2O + h\upsilon \rightarrow Fe^{2+} + OH^{\bullet} + H^+ \qquad (51)$$

ANN modeling of photo–Fenton processes is the subject of numerous papers (see Table 7). Durán et al [192] have developed a three layers (4:4:2) feed–forward network to predict degradation rate constant of cyanides and formates under UV/Fe(II)/H_2O_2 process in a integrated gasification combined cycle (IGCC) power station effluent. In a similar work, they have used a two layers network to evaluate the efficiency of photo–Fenton process [193]. The network has been trained by a marquardt non–linear fitting algorithm to simulate the output parameters: decolorization and mineralization rate constants. Simulation from ANNs equations has proved that the initial concentration of hydrogen peroxide in aqueous dye solutions is the main parameter affecting the photo–decolorization kinetics (See Table 7).

Gob et al. [181] have studied photo–Fenton removal of 2,4–dimethyl aniline (2,4–xylidine) from contaminated water. A three layers (3:8:1) feed–forward back propagation network has been developed and trained using 50000 data sets. Comparison made between predicted and experimental output values (R^2=0.995) show that ANN is a successful technique to predict 2,4–xylidine concentration in the treated solution (See Table 7).

Treatment of saline wastewater contaminated with hydrocarbons by the photo–Fenton process has been the subject of another ANN modeling. In this work, Moraes et al. [8] have followed TOC content of the treated wastewater using a three layer (5:2:1) feed–forward back propagation network. Totally 1000 data sets have been used for training the network. There has been a good agreement between experimental and predicted output values with high correlation coefficients of 0.950 and 0.965 for learning and test sets, respectively.

Monteagudo et al. [194] have also developed a neural network for modeling of ferrioxalate–assisted solar photo–Fenton degradation of Orange II aqueous solutions. The three layers (7:2:2) feed–forward back propagation network has been trained using Marquardt non–linear fitting algorithm which can predict the decolorization kinetic rate constant chosen as the output variable. They have reported that experimental results and ANNs fittings of the process are in good agreement with an average error lower than 16% for dye decolorization.

Calza et al. [152] have developed a neural network to predict the performance of a photo–Fenton process for removal of imipramine from the contaminated water. The three layers network (3:3:1) has been trained with conjugate gradient descent algorithm during 95 epochs. The linear regression between the network prediction and the corresponding experimental data prove that modeling of photo–Fenton removal of imipramine using artificial neuron network is a satisfactory method. Modeling photo–Fenton removal of Reactive Blue 4 is another example of the application of neural network technique. Duran et al. [154] have developed a three layers (4:2:1) feed–forward neural network using back propagation algorithm. Totally 19 data sets and marquardt non–linear fitting algorithm have been used for training the network that is successfully enabled to predict the output variable, i.e. decolorization kinetic constant.

Giroto and co–workers [195] have also developed a neural network to predict polyvinyl alcohol abatement in aqueous solution by photo–Fenton process. A three layers (4:8:1) feed–forward back propagation network has been trained using 432 data sets and during 10000 epochs to predict polyvinyl alcohol concentration at the end of photo–Fenton process. High correlation coefficient (R^2=0.996) between experimental and predicted values of the output variable shows the success of the modeling.

Nogueira et al. [196] have also developed a neural network for modeling the performance of a solar driven photo–Fenton process used for removal of phenol from the effluents. In this work, dissolved organic carbon (DOC) content of the treated solution has been chosen as the output variable and a three layers (5:6:1) feed-forward back propagation neural network has been trained during 10000 epochs.

Homogeneous photo–Fenton has some disadvantages such as (i) the tight range of pH in which the reaction proceeds, (ii) the need for recovering the precipitated catalyst after the treatment and (iii) deactivation by some ion complexing agents like phosphate anions [197]. An alternative method could be the use of heterogeneous solid Fenton catalysts, such as transition metals containing zeolites, clays, bentonits and so on. [198–200]. The use of synthetic zeolites is very promising due to their unique properties such as micro–porous structure, high surface area and ion exchange capacity, which could give them advantage over other carriers [201–203].

Kasiri et al. [24] have studied photo–Fenton process using Fe–ZSM5 zeolite as heterogeneous catalyst for removal of Acid Red 14 from the contaminated water. They have used an artificial neural network for modeling of the process. The three layers (4:10:1) network has been trained using scaled conjugate gradient algorithm and the degradation efficiency of the process has been chosen as the output variable. Modeling results show a good agreement between experimental and predicted results with a high correlation coefficient (R^2=0.996). The results of modeling confirm that neural network modeling could effectively reproduce experimental data and predict the behavior of the process.

9.2. UV/H_2O_2, OZONATION AND CHLORINATION PROCESSES

The principle behind the beneficial effects observed using ultraviolet light in combination with hydrogen peroxide or ozone as compared to the individual application, lies in the fact that the rate of generation of free radicals is significantly enhanced in the case of combination technique, which is very similar to ultrasoundy H_2O_2 or ultrasoundy O_3 processes. Only difference being the energy required for the generation of free radicals from dissociation of ozone or hydrogen peroxide is given by the UV light as against cavitating bubbles in the case of ultrasound [207, 208].

It is widely accepted that the first step in the UV/H_2O_2 process is the attack of the photon against the hydrogen peroxide molecule and the subsequent formation of hydroxyl radical (OH$^•$) (Eq. (31)) [209].

High concentrations of H_2O_2 do not necessarily favor the kinetics of the reaction, for after the reaction starts, the steps of propagation can be prevented by the excess of hydrogen peroxide. This excess can act as a hydroxyl radical self–consumer [210], according to the reaction (37).

Besides water, reaction (37) produces the hydroperoxy radical, less reactive than the hydroxyl radical. Thus, hydrogen peroxide in excess may react with the hydroxyl radical and compete with the attack of this radical to the organic matters in the solution during the photolysis [211].

Table 7. ANN modeling of Fenton and photo–Fenton water and wastewater treatment processes

Treatment Process	Treatment Target	ANN Architecture	Training Function	Layers No.	ANN Topology	Data No.	Input	Output	Epochs No.	Ref.
Photo–Fenton	Saline solutions containing raw gasoline	Feed–forward back propagation	–	3	5:2:1	–	Reaction time, $[TOC]_0$, $[Fe(II)]_0$, $[H_2O_2]_0$, $[NaCl]_0$	*[TOC]	1000	[8]
UV/ Fe–ZSM5/ H_2O_2	Acid Red 14	Feed–forward back propagation	Scaled conjugate gradient algorithm	3	4:10:1	25	$[Dye]_0$, $[H_2O_2]_0$, Initial pH, $[Catalyst]_0$	Degradation (%)	–	[24]
Photo–Fenton	Reactive Blue 4	Feed–forward back propagation	Marquardt non–linear fitting algorithm	3	4:2:1	19	$[H_2O_2]_0$, pH, $[TiO_2]_0$, $[Fe(II)]_0$	Decolorization kinetic constant	–	[132]
Photo–Fenton	Imipramine	Feed–forward back propagation	Conjugate gradient descent	3	3:3:1	–	$[H_2O_2]_0$, $[Fe(II)]_0$, $[TiO_2]_0$	Imipramine degradation (%)	95	[152]
Photo–Fenton	2,4-dimethyl aniline	Feed–forward back propagation	–	3	3:8:1	–	$[Fe(II)]_0$, $[H_2O_2]$, Reaction time	*[Xyl]	50000	[181]

* **TOC**: Total Organic Carbon, **Xyl**: 2,4-dimethyl aniline.

Treatment Process	Treatment Target	ANN Architecture	Training Function	Layers No.	ANN Topology	Data No.	Input	Output	Epochs No.	Ref.
Fenton	Antibiotics (amoxicillin, ampicillin and cloxacillin)	Feed–forward back propagation	Levenberg–Marquardt	3	5:14:1	120	Reaction time, H_2O_2/COD molar ratio, $H_2O_2/Fe(II)$ molar ratio, pH, $[Antibiotics]_0$	*COD removal (%)	–	[184]
Fenton	Reactive Brilliant Blue, Reactive Blue 49	–	–	–	–	–	$[Fe(II)]_0$, $[H_2O_2]_0$	COD removal, Effluent color removal	–	[185]
UV/Fe(II)/ H_2O_2	Integrated Gasification Combined Cycle (IGCC) power station effluent	Feed–forward back propagation	–	3	4:2:2	–	$[Fe(II)]_0$, $[H_2O_2]_0$, pH, Temperature	Degradation rate constant of cyanides and formates	–	[192]
Photo–Fenton	Reactive Blue 4 (RB4)	Marquardt non–linear fitting algorithm	–	2	5:2	–	$[RB4]_0$, $[Fe(II)]_0$, $[H_2O_2]_0$, pH, Temperature	Decolorization and mineralization rate constant	–	[193]

* **COD**: Chemical Oxygen Demand.

Table 7. (Continued).

Treatment Process	Treatment Target	ANN Architecture	Training Function	Layers No.	ANN Topology	Data No.	Input	Output	Epochs No.	Ref.
Ferrioxalate-assisted photo-Fenton	Orange II	Feed–forward back propagation	Marquardt non–linear fitting algorithm	3	7:2:2	–	H_2O_2 Flow, $[Fe(II)]_0$, pH, $[H_2C_2O_4]_0$, Temperature Solar power, UV Dosage	Decolorization and mineralization kinetic rate constants	–	[194]
Photo-Fenton	Polyvinyl alcohol (PVA)	Feed–forward back propagation	–	3	4:8:1	432	Reaction time, $[DOC]_0$, $[Fe(II)]_0$, $[H_2O_2]$	[PVA]	10000	[195]
Solar driven photo-Fenton	Phenol	Feed–forward back propagation	–	3	5:6:1	–	$[DOC]_0, [Fe(II)]_0$, H_2O_2 feed rate, Accumulated radiant energy, Irradiation time	*[DOC]	10000	[197]
Solar photo-Fenton	Reactive Blue 4 (RB4)	Feed–forward back propagation	Marquardt non–linear fitting algorithm	3	5:2:1	–	pH, $[Fe(II)]_0$, $[H_2O_2]_0$, $[RB4]_0$, Temperature	Decolorization rate pseudo constant	–	[204]
Solar photo-Fenton–ferrioxalate					6:2:1		$[Oxalic\ acid]_0$			

* **DOC**: Dissolved Organic Carbon.

Table 7. (Continued).

Treatment Process	Treatment Target	ANN Architecture	Training Function	Layers No.	ANN Topology	Data No.	Input	Output	Epochs No.	Ref.
Solar photo–Fenton	Orange II	–	–	3	5:2:1	–	$[H_2O_2]_0$, $[FeSO_4]_0$, [Orange II]$_0$, pH, Temperature	Decolorization kinetic rate constant	–	[205]
Solar photo–Fenton–ferrioxalate					4:2:1		$[FeSO_4]_0$, $[H_2O_2]_0$, Temperature, [Oxalic acid]$_0$			
Ferrioxalate-assisted solar photo–Fenton	Orange II	–	–	3	7:2:2	50	pH, [Fe(II)]$_0$, H_2O_2 flow, Temperature, $[H_2C_2O_4]_0$, Solar power, Air flow	Decolorization and mineralization kinetic rate constants	–	[206]

Considering the recombination reactions (Eqs. (52-55)), there is the possibility of hydroxyl radical consumption and consequently, the probability of decreasing oxidation of the organic matters. Thus, a competition for the ultraviolet light starts.

$$H_2O_2 + HO_2^\bullet \rightarrow OH^\bullet + H_2O + O_2 \tag{52}$$

$$OH^\bullet + OH^\bullet \rightarrow H_2O_2 \tag{53}$$

$$HO_2^\bullet + HO_2^\bullet \rightarrow H_2O_2 + O_2 \tag{54}$$

$$OH^\bullet + HO_2^\bullet \rightarrow H_2O + O_2 \tag{55}$$

The kinetics of UV/H_2O_2 process is favored up to H_2O_2 addition critical point. The critical point is related to various factors such as the amount of hydrogen peroxide added, reaction media pH, UV radiation wavelength, concentration of organic matters and structural characteristics, besides other specific factors like the presence of inorganic salts, which affect the reaction performance of the hydroxyl radical [212].

Wastewater treatment by applying photooxidative processes is, in general, quite complex. The mathematical equations describing the performance of theses processes involve the radiant energy balance, the spatial distribution of the absorbed radiation, mass transfer, and the mechanisms involving radical species. It is clear that such a complex problem cannot be solved by simple linear multivariate correlation. Artificial neural networks are very useful to solve this problem as they do not require the mathematical description of the phenomena involved in the process [23, 65, 213, 214]. So, there are numerous papers regarding ANN modeling of these processes (see Table 8).

Aleboyeh et al. [23] have developed an artificial neural network to predict the performance of a UV/H_2O_2 removal of Acid Orange 7 from the aqueous solution. The network has been trained using totally 228 data sets divided into training, validation and test subsets, each of them containing 114, 57 and 57 data sets, respectively. A three layers network (4:8:1) has been optimized to predict the decolorization efficiency of UV/H_2O_2 process. The comparison between experimental values and predicted output variables using the adopted neural network model show that this network predicts the output variable with a high correlation coefficient (R^2=0.996). The results of modeling confirm that neural network modeling could effectively reproduce experimental data and predict the behavior of the process.

Salari and co-workers [215] have used ANN technique for modeling of MTBE removal by UV/H_2O_2 process. Totally 64 data sets have been used for training, validation and test of the model. The configuration of the back propagation neural network giving the smallest MSE has been a three layers ANN with tangent sigmoid transfer function (tansig) at hidden layer with 8 neurons, linear transfer function (purelin) at output layer trained with scaled conjugate gradient algorithm. ANN predicted results are very close to the experimental ones with correlation coefficient of 0.998. The ANN model can then describe the behavior of the complex reaction system with the range of experimental conditions adopted.

Modeling of UV/H_2O_2 removal of Reactive Red 120 has been also studied by Slokar et al. [25]. A four layers (7:8:8:4) neural network has been developed during 200 epochs on the basis of counter–propagation learning strategy and Kohonen algorithm. This optimized network can effectively predict the output variables including absorbance, chemical oxygen demand, total organic carbon and total inorganic carbon of the dye solution (See Table 8).

Guimaraes and Silva have established a hybrid neural model for decolorization of azo dyes by UV/H_2O_2 involving the study of process variables and structural parameters [216]. Decolorization degree of the studied dyes including Direct Red 28, Acid Brown 75, Acid Orange 52 and Orange 10 has been chosen as the output variable. The network has been trained by Levenberg–Marquardt algorithm during 34 epochs and consequently, a three layers (7:18:1) neural network is optimized for modeling of UV/H_2O_2 process. The neural model provided optimum estimates for the decolorization based on the absorbance measurement as an output variable, with correlation coefficients above 0.96 for the training, validation and test sets, indicating the optimum model generalization capacity.

Ozone has been demonstrated to oxidize a variety of inorganic matters, humic substances (HS) and toxic contaminants found in drinking water [217]. Thus, ozonation method is preferred in many instances for treating drinking water to achieve both disinfection and oxidation [217]. The ozone oxidation is accomplished in two pathways: direct oxidation by molecular ozone and indirect oxidation by hydroxyl radical [218]. The direct ozone oxidation reactions are highly selective but relatively slow attacking the unsaturated electron–rich bonds contained in specific functional groups, aromatics, olefins and amines [219]. In comparison, the indirect reaction has a relatively low selectivity but a quick reaction rate by hydroxyl radicals, which are generated by decomposition of ozone molecule [220, 221]. The hydroxyl radicals can oxidize regular organic substrates, microorganisms and NH_3–nitrogen.

Ozonation is a complex process involving different phenomena. More complex situation arises when we deal with unknown parameters as well as incomplete state observations. Therefore, neural network modeling has been suggested to be applied to provide a good enough state estimates (e.g. the organic compounds elimination from water by ozonation), without either preliminary parameter identification, or exact model structure knowledge (See Table 8) [222].

ANN modeling of the removal of humic substances from the aqueous solutions with ozonation has been presented by Oguz et al. [223]. Simulation has been done on the basis of Levenberg–Marquardt algorithm and a three layers (8:10:1) feed–forward network is developed to predict humic substances percentage as the output variable. A relationship between the predicted results of the designed ANN model and experimental data has been also conducted. As a result of ANN model, the values of correlation coefficient, standard deviation ratio, mean absolute error and root mean square error have been obtained as 0.995, 0.065, 4.057 and 5.496, respectively.

In a similar work, Oguz and co–workers have developed a neural network to predict the apparent ozonation rate constants of 1:2 metal complex dyestuffs [224]. They have proposed a three layers (8:10:1) feed–forward back propagation network that has been optimized using 203 data sets and Levenberg–Marquardt algorithm. The model based on artificial neural network can predict the concentrations of dyestuffs removal from the aqueous solution during ozonation under the different conditions.

Legube et al. [225] have also studied ANN modeling of bromate formation by ozonation of surface waters in drinking water treatment. A three layers (7:3:1) multilayer perceptron network has been trained using 204 data sets and conjugate gradient algorithm to predict bromate concentration at the end of ozonation process. High correlation coefficient (R^2=0.98) between experimental and predicted values of residual bromate proven the success of the modeling.

Chlorination/dechlorination is the most common process used for wastewater disinfection [226]. Total coliform count in wastewater effluent is a primary factor for wastewater reuse, and is affected by chlorination dose. However, ammonia and organic–N in influent can react with chlorine to form chloramines and organochloramines [227–229]. Additionally, total coliform counts cannot be monitored on–line, thereby increasing the difficulty in controlling chlorination dosage. Conversely, chlorinated effluent always contains significant chlorine residuals that are disadvantageous constituent for the receiving water body. Therefore, dechlorination is always required. Controlling chlorine residuals in effluent is complex and dynamic, and varies depending on the influent ammonia concentration, chlorine and dechlorination dosages [230]. However, ANN modeling as a powerful modeling technique has been successfully used for controlling the chlorine residuals.

Yu et al. [230] have developed a neural network as a dynamic control of disinfection by chlorination for wastewater reuse. Totally 411 data sets and generalized delta–learning rule have been used for simulation of three layers networks to predict different chemical and physico–chemical parameters of the treated real wastewater (city sewer). Experimental results show that chlorination and dechlorination doses can be effectively controlled using the ANN control models.

As a conclusion, artificial neural networks can successfully describe the performance of the photooxidative processes in the range of the variables studied, in spite of the complexity involved such processes.

Table 8. ANN modeling of UV/H_2O_2, ozonation and chlorination processes for water and wastewater treatment

Treatment Process	Treatment Target	ANN Architecture	Training Function	Layers No.	ANN Topology	Data No.	Input	Output	Epochs No.	Ref.
UV/H_2O_2	Acid Orange 7	Feed-forward back propagation	Scaled conjugate gradient algorithm	3	4:8:1	228	[Dye]$_0$, [H_2O_2]$_0$, Initial pH, Reaction time	Color removal (%)	–	[23]
UV/H_2O_2	Reactive Red 120	Counter-propagation learning strategy	Kohonen algorithm	4	7:8:8:4	–	UV light intensity, [H_2O_2]$_0$, [NaCl]$_0$, Reaction time, [NaOH]$_0$, [Dye], [*Oxidant]$_0$	Absorbance, *COD, TOC, TIC	200	[25]
UV/peroxy-disulfate	Basic Blue 3	Feed-forward back propagation	Scaled conjugate gradient algorithm	3	4:8:1	177	Reaction time, [Dye]$_0$, [Peroxydisulfate]$_0$, UV light intensity,	Color removal (%)	5000	[27]
UV/H_2O_2	Methyl tert-butyl etherMTBE)	Feed-forward back propagation	Scaled conjugate gradient algorithm	3	4:8:1	64	[MTBE]$_0$, [H_2O_2]$_0$, pH, Reaction time	[MTBE]$_t$	–	[215]
UV/H_2O_2	Acid Brown 75, Acid Orange 52, Acid Orange 10, Direct Red 28	Feed-forward back propagation	Levenberg–Marquardt	3	7:18:1	757	Azo bond number, Sulphonate group number, [Dye]$_0$, pH, H_2O_2 volume, Temperature, Time of operation	Decolorization degree	34	[216]

* **COD**: Chemical Oxygen Demand; **TOC**: Total Organic Carbon; **TIC**: Total Inorganic Carbon,

Table 8. (Continued).

Treatment Process	Treatment Target	ANN Architecture	Training Function	Layers No.	ANN Topology	Data No.	Input	Output	Epochs No.	Ref.
Ozonation	Phenols (phenol (PH), 4-chlorophenol (4-CPH) and 2,4-dichlorophenol (2,4-DCPH))	Dynamic neuro network observer (DNNO)	Modified version of the least square algorithm	3	–	–	$[Ozone]_{gas}$, Organic compound concentration, Volume of gas phase, Gas flow rate, Reaction constant, Liquid phase volume	Organics decomposition dynamics of phenols and their mixtures	–	[222]
Ozonation	Humic substances (HS)	Feed–forward back propagation	Levenberg–Marquardt algorithm	3	8:10:1	–	Treatment time, $[HS]_0$, pH, *PAC, Ozone–air flow rate, $[HCO_3^-]_0$, Ozone generation potential, Temperature	HS removal (%)	–	[223]
Ozonation	Metal complex dyestuffs	Feed–forward back propagation	Levenberg–Marquardt algorithm	3	8:10:1	203	Treatment time, $[HCO_3^-]_0$, Temperature, PAC, Ozone–air flow rate, pH, O_3 percentages, Air flow rate, $[H_2O_2]_0$	Dyestuffs removal (%)	–	[224]

* **PAC**: Powdered Activated Carbon.

Table 8. (Continued).

Treatment Process	Treatment Target	ANN Architecture	Training Function	Layers No.	ANN Topology	Data No.	Input	Output	Epochs No.	Ref.
Ozonation	Bromate formation	Multilayer–perceptron (MLP)	Conjugate gradient	3	7:3:1	204	pH, Alkalinity, Temperature, $[Br^-]_0$, $[N–NH_4^+]$, DOC, UV light intensity	$[BrO_3^-]$	–	[225]
Chlorination	Real wastewater (City sewer)	Feed–forward back propagation	Generalized delta–learning rule	3	5:9:1	411	pH/*ORP values in the pre–mixing tank and chlorination reactor, Chlorine dose, Dechlorination dose	Chlorine dose, Effluent total coliform count, Dechlorination dose, Effluent chlorine residuals, Chlorine residuals in chlorinated Reactors	5300	[230]
					5:20:1				4800	
					8:14:1				1500	
					8:20:1				2400	
					5:10:1				9300	
UV/H_2O_2	Acid Orange 52	Feed–forward back propagation	Levenberg–Marquardt	3	5:16:1	218	H_2O_2 volume, pH, Temperature, $[Dye]_0$, Time of operation	Decolorization (%)	22	[231]
UV/H_2O_2	Acid Brown 75	Feed–forward back propagation	Levenberg–Marquardt	3	5:116:1	528	$[Dye]_0$, pH, H_2O_2 volume, Temperature, Time of operation	Decolorization degree	49	[232]

* **ORP**: Oxidation-Reduction Potential.

CONCLUSIONS

Due to the complexity of water and wastewater treatment processes, the effect of different parameters involved are very difficult to determine, leading to uncertainties in the design and scale-up of industrial interest. Artificial neural networks are powerful modeling techniques that do not require the mathematical description of the phenomena involved in the process. ANNs are capable to simulate the complex relationships existing between input and output process variables in different treatment methods. This technique has been widely used in modeling and simulation of electrochemical, photocatalytic, biological, photooxidative and adsorption processes. The results of this review showed that neural network modeling is an effective and simple approach to successfully describe the behavior of complex water and wastewater treatment processes, in which manipulated operational variables show a combined effect, within the range of experimental conditions investigated. Comparison between the predicted results of the designed ANN model and experimental data show a high correlation coefficient in almost all the cases. The linear regression between the network prediction and the corresponding experimental data also prove that modeling these processes using artificial neuron network is a satisfactory method. This technique might therefore be useful in process optimization, as well as in the design, scale-up and industrial application of water and wastewater treatment processes.

Real effort has been put to include in the review of all the relevant referred publications on the application of artificial neural networks modeling for water and wastewater treatment processes. However, the limitation of our resources and the sheer number of publications in this field may have prevented the comprehensiveness of this report. Our sincere apologies are extended to any and all authors whose works are not included in this report.

ACKNOWLEDGMENTS

We are grateful to the University of Tabriz and Tabriz Islamic Art University, Iran for the all supports.

REFERENCES

[1] McCulloch, WS; Pitts, W. *Bull. Mafh. Biophysics,* 1943, 5, 113-133.
[2] Krose, B; Smagt, PV. *An Introduction to Neural Networks*, The University of Amsterdam, Eighth edition, 1996.
[3] Zhou, H; Smith, DW. *J. Environ. Eng. Sci.*, 2002, 1, 247-264.
[4] Mallevialle, J; Odendall, PE; Wiesner, MR. *Water Treatment Membrane Processes*, McGraw-Hill, NewYork, 1996.
[5] Despange, F; Massart, DL. *Analyst,* 1998, 123, 157-178.
[6] Lek, S; Guegan, JF. *Ecolo. Model.,* 1999, 120, 65-73.
[7] Pareek, VK; Brungs, MP; Adesina, AA; Sharma, R. *J. Photochem. Photobiol., A,* 2002, 149, 139-146.
[8] Moraes, JEF; Quina, FH; Nascimeto, CAO; Silva, DN; Chiavone-Filho, O. *Environ. Sci. Technol.,* 2004, 38, 1183-1187.
[9] Zarei, M; Niaei, A; Salari, D; Khataee, AR. *J. Electroanal. Chem.,* 2010, 639, 167-174.
[10] Howard, D; Beale, M; Hagan, M; *Neural Network Toolbox For Use with MATLAB® User's Guide*, Version 5, Mathworks Inc, 2006.
[11] Moller, MF. *Neural Networks,* 1993, 6, 525-533.
[12] Zhang, Z; Friedrich, K. *Compo. Sci. Technol.,* 2003, 63, 2029-2044.
[13] Parizeau, M. Réseaux de neurones, GIF-21140 et GIF-64326, *Université Laval*, 2006.
[14] Si, J; Barato, AG; Powell, WB; Wunsch, D; *Handbook of learning and approximate dynamic programming*, IEEE press, John Wiley & Sons, 2004.
[15] Robert, C; Guilpin, C; Limoge, A. *J. Neurosci. Methods,* 1998, 79, 187-193.

[16] Gob, S; Oliveros, E; Bossmann, SH; Braun, AM; Guardani, R; Nascimento, CAO. *Comput. Ind. Eng.*, 2007, 53, 95-122.
[17] Bishop, C. *Neural Networks for Pattern Recognition*, Oxford University Press, USA, 1995.
[18] Alvarez, A. *Neural Process. Lett.*, 2002, 16, 43-52.
[19] Neudecker, H; Magnus, JR. Matrix differential calculus with applications in statistics and econometrics, New York, John Wiley & Sons, 1998, 136
[20] Brinhole, ER; Destro, JFZ; de Freitas, AAC; de Alcantara Jr., NP. *Progress In Electromagnetics Research Symposium*, Hangzhou, China, 2005, August 22-26.
[21] Hamed, MM; Khalafallah, MG; Hassanien, EA. *Environ. Model, Soft.*, 2004, 19, 919-928.
[22] Salari, D; Niaei, A; Khataee, AR; Zarei, M. *J. Electroanal. Chem.*, 2009, 629, 117-125.
[23] Aleboyeh, A; Kasiri, MB; Olya, ME; Aleboyeh, H. *Dyes Pigments*, 2008, 77, 288-294.
[24] Kasiri, MB; Aleboyeh, H; Aleboyeh, A. *Environ. Sci. Technol.*, 2008, 42, 7970-7975.
[25] Slokar, YM; Zupan, J. Marechal, AML; *Dyes Pigments,* 1999, 42, 123-135.
[26] Garson, GD. *AI Expert*, 1991, 6, 47-51.
[27] Khataee, AR; Mirzajani, O. *Desalination*, 2010, 251, 64-6.
[28] Wong, BK; Selvi, Y. *Inform. Manage.*, 1998, 34, 129-139.
[29] Schocken, S; Ariav, G. *Decis. Support Sys.,* 1994, 11, 393-414.
[30] Constantin Zopounidis, C; Doumpos, M. *Eur. J. Oper. Res.*, 2002, 139, 371-389.
[31] Dijk, DV. *Int. J. Forecasting*, 2006, 22, 407-408.
[32] O'Connor, N; Madden, MG. *Knowl-Based Syst.*, 2006, 19, 371-378.
[33] Kaastra, I; Boyd, M. *Neurocomputing*, 1996, 10, 215-236.
[34] Wong, BK; Bodnovich, TA; Selvi, Y. *Decis. Support Syst.*, 1997, 19, 301-320.
[35] Paliwal, M; Kumar, UA. *Expert Syst. Appl.*, 2009, 36, 2-17.
[36] Kaefer, F; Heilman, CM; Ramenofsky, SD. *Comput. Oper. Res.*, 2005, 32, 2595-2615.
[37] Chiu, CY; Chen, YF; Kuo, IT; Ku, HC. *Expert Syst. Appl.*, 2009, 36, 4558-4565.
[38] Hussain, MA. *Artif. Intell. Eng.*, 1999, 13, 55-68.
[39] Lisboa, PJG; Taktak, AFG. *Neural Networks*, 2006, 19, 408-415.

[40] Lisboa, PJG. *Neural Networks*, 2002, 15, 11-39.
[41] Lisboa, PJG; Etchells, TA; Jarman, IH; Aung, MSH; Chabaud, S; Bachelot, T; Perol, D; Gargi, T; Bourdès, V; Bonnevay, S; Négrier, S. *Neural Networks*, 2008, 21, 414-426.
[42] Baxt, WG. *Lancet*, 1995, 346, 1135-1138.
[43] Reggia, JA. *Artif. Intell. Eng.*, 1993, 5, 143-157.
[44] Khanmohammadi, M; Garmarudi, MB; Khoddami, N; Shabani, K; Khanlari, M. *Microchem. J.*, 2010, 95, 337-340.
[45] Rashidi, AM; Eivani, AR; Amadeh, A. *Comp. Mater. Sci.*, 2009, 45, 499-504.
[46] Gaganis, C; Pasiouras, F; Doump, M. *Expert Syst. Appl.*, 2007, 32, 114-124.
[47] Bjornson, C; Barney, DKB. *Expert Syst. Appl.*, 1999, 17, 13-19.
[48] Leung, YW; Mao, JY. *Expert Syst. Appl.*, 2003, 25, 255-267.
[49] Hruschka, H. *Eur. J. Oper. Res.*, 1993, 66, 27-35.
[50] Vellido, A; Lisboa, PJG; Meehan, K. *Expert Syst. Appl.*, 1999, 17, 303-314.
[51] Gareta, R; Romeo, LM; Gil, A. *Energ. Convers. Manage.*, 2006, 47, 1770-1778.
[52] Guven, A; Sogukpinar, I. *Comput. Secur.*, 2003, 22, 695-706.
[53] Doumas, A; Mavroudakis, K; Gritzalis, D; Katsikas, S. *Comput. Secur.*, 1995, 14, 435-448.
[54] Arai, K. Proceedings in Marine Science, 2008, 9, 43-51.
[55] Akyol, DE; Bayhan, GM. *Comput. Ind. Eng.*, 2007, 53, 95-122.
[56] Fernando Morgado Dias, FM; Antunes, A; Mota; AM. *Eng. Appl. Artif. Intel.*, 2004, 17, 945-952.
[57] Marini, F. *Anal. Chim. Acta*, 2009, 635, 121-131.
[58] Cramer, C. *Eur. J. Oper. Res.*, 1998, 108, 266-282.
[59] Egmont-Petersen, M; de Ridder, D; Handels, H. *Pattern Recogn.*, 2002, 35, 2279-2301.
[60] Jiang, J. *Signal Process. Image Commun.*, 1999, 14, 737-760.
[61] Koivisto, H; Ruoppila, V; Koivo, HN. *Ann. R. Aut. P.*, 1992, 17, 61-6.
[62] Anagnostou, T; Remzi, M; Lykourinas, M; Djavan, B. *Eur. Urol.*, 2003, 43, 596-603.
[63] Heijst, JJV; Touwen, BCL; Vos, JE. *Early Hum. Dev.*, 1999, 55, 77-95.
[64] Fogel, DB; Wasson III, EC; Boughton, EM. *Cancer Lett.*, 1995, 96, 49-53.
[65] Niaie, A; Towfighi, J; Khataee, AR; Rostamizadeh, K. *Pet. Sci. Technol.*, 2007, 25, 967-982.

[66] Mingzhi, H; Jinquan, W; Yongwen, M; Yan, W; Weijiang, L; Xiaofei, S. *Expert Syst. Appl.*, 2009, 36, 10428-10437.
[67] Huang, W; Foo, S. *Water Res.*, 2002, 36, 356-362.
[68] Wei, B; Sugiura, N; Maekawa, T. *Water Res.*, 2001, 35, 2022-2028.
[69] Kurt, A; Gulbagci, B; Karaca, F; Alagha, O. *Environ. Int.*, 2008, 34, 592-598.
[70] Maier, HR; Dandy, GC. *Math. Comput. Model.*, 2001, 33, 669-682.
[71] Emilio, CA; Magallanes, JF; Litter, MI. *Anal. Chim. Acta*, 2007, 595, 89-97.
[72] Levine, ER; Kimes, DS; Sigillito, VG. *Ecol. Model.*, 1996, 92, 101-108;
[73] Haciismailoglu, MC; Kucuk, I; Derebasi, N. *Expert Syst. Appl.*, 2009, 36, 2225-2227.
[74] Raymond, JW; Rogers, TN; Shonnard, DR; Kline, AA. *J. Hazard. Mater.*, 2001, 84, 189-215.
[75] Ren, Q; Cao, Q. *T. Nonfer. Metal. Soc. China*, 2006, 16, 865-868.
[76] Kilmer, RA. *Math. Comput. Model.*, 1996, 23, 91-99.
[77] Erkmen, B; Yıldırım, T. *Expert Syst. Appl.*, 2008, 35, 472-475.
[78] Dov Dvir, D; Ben-David, A; Sadeh, A, Shenhar, AJ. *Eng. Appl. Artif. Intel.*, 2006, 19, 535-543.
[79] Rodrigue, JP. *Transport. Res. C-Emer.*, 1997, 5, 259-271.
[80] Dharia, A; Adeli, H. *Eng. Appl. Artif. Intel.*, 2003, 16, 607-613.
[81] Celikoglu, HB; Cigizoglu, HK. *Math. Comput. Model.*, 2007, 45, 480-489.
[82] Berke, L; Patnaik, SN; Murthy, PLN. *Comput. Struct.*, 1993, 48, 1001-1010.
[83] Savran, A; Tasaltin, R; Becerikli, Y. *ISA Transactions*, 2006, 45, 225-247.
[84] Melin, P; Castillo, O. *Appl. Soft Comput.*, 2003, 3, 353-362.
[85] Li, Y; Sundararajan, N; Saratchandran, P. *Automatica*, 2001, 37, 1293-1301.
[86] Lin, SH; Juang, RS. *J. Environ. Manage.*, 2009, 90, 1336-1349.
[87] Rafatullah, M; Sulaiman, O; Hashim, R; Ahmad, A. *J. Hazard. Mater.*, 2010, 177, 70-80.
[88] Daneshvar, N; Aber, S; Khani, A; Khataee, AR. *J. Hazard. Mater.*, 2007, 144, 47-51.
[89] Kedziorek, MAM; Bourg, ACM; Giffaut, E. *Phys. Chem. Earth Parts A/B/C*, 2007, 32, 568-572.
[90] Jiao, XC; Xu, FL; Dawson, R; Chen, SH; Tao, S. *Environ. Pollut.*, 2007, 148, 230-235.

[91] Khataee, AR; Khani, A. *Int. J. Chem. React. Eng.*, 2009, 7, Article 5.
[92] Aber, S; Daneshvar, N; Soroureddin, SM; Chabok, A; Asadpour-Zeynali, K. *Desalination*, 2007, 211, 87-95.
[93] Yetilmezsoy, K; Demirel, S. *J. Hazard. Mater.*, 2008, 153, 1288-1300.
[94] Faur, C; Cougnaud, A; Dreyfus, G; Le Cloirec, P. *Chem. Eng. J.*, 2008, 145, 7-15.
[95] Brasquet, C; Le Cloirem P. *Water Res.*, 1999, 33, 3603-3608.
[96] Fu, RQ; Xu, TW; Pan, ZX. *J. Membrane Sci.*, 2005, 251, 137-144.
[97] Leslie Grady, CP; Daigger, GT; Lim; HC. *Biological wastewater treatment*, Second Edition, Marcel Dekker; Inc. 1999.
[98] Banat, IM; Nigam, P; Singh, D. *Bioresource Technol.*, 1996, 58, 217-227.
[99] JS. Chang, JS; Kua, TS; Chao, YP; Ho, JY; Lin, PJ. *Biotechnol. Lett.*, 2000, 22, 807-812.
[100] Chang, JS; Kuo, TS. *Bioresour. Technol.*, 2000, 75, 107-111.
[101] Fu, Y; Viraraghavan, T. *Bioresource Technol.*, 2002, 82, 139-145.
[102] Khataee, AR; Zarei, M; Pourhassan, M. *Environ. Technol.*, 2009, 30, 1615-1623.
[103] Khataee, AR; Zarei, M; Pourhassan, M. *Clean- Soil, Air, Water*, 2010, 38, 96-103.
[104] Khataee, AR; Dehghan, G; Ebadi, E; Zarei, M; Pourhassan, M. *Bioresource Technol.*, 2010, 101, 2252-2258.
[105] Khataee, AR; Dehghan, G; Zarei, M; Ebadi, E; Pourhassan, M. *Chem. Eng. Res. Des.*, 2010, Inpress, Doi:10.1016/j.cherd.2010.05.009.
[106] Prakash, N; Manikandan, SA; Govindarajan, L; Vijayagopal, V. *J. Hazard. Mater.*, 2008, 152, 1268-1275.
[107] Venu Vinod, A; Arun Kumar, K; Venkat Reddy, G. *Biochem. Eng. J.*, 2009, 46, 12-20.
[108] Arranz, A; Bordel, S; Villaverde, S; Zamarreno, JM; Guieysse, B; Munoz, R. *J. Hazard. Mater.*, 2008, 155, 51-57.
[109] Fagundes-Klen, MR; Ferri, P; Martins, TD; Tavares, CRG; Silva, EA. *Biochem. Eng. J.*, 2007, 34, 136-146.
[110] Yang, H; Jiang, Z; Shi, S. *Sci.the Total Environ.*, 2006, 358, 265- 276.
[111] Mjalli, FS; Al-Asheh, S; Alfadala, HE. *J. Environ. Manag.*, 2007, 83, 329-338.
[112] Sahinkaya, E. *J. Hazard. Mater.*, 2009, 164, 105-113.
[113] Basu, B; Singh, MP; Kapur, GS; Ali, N; Sastry, MIS; Jain, SK; Srivastava SP; Bhatnagar, AK. *Tribol. Int.*, 1998, 31, 159-168.
[114] Nagy, ZK. *Chem. Eng. J.*, 2007, 127, 95-109.

[115] Hong, SH; Lee, MW; Lee, DS; Park, JM. *Biochem. Eng. J.*, 2007, 35, 365-37.
[116] Cinar, O; Hasar, H; Kinaci, C. *J. Biotech.*, 2006, 123, 204-209.
[117] Rajeshwar, K; Ibanez, JG. Environmental Electrochemistry Fundamentals and Applications in Pollution Abatement, Academic Press, San Diego, CA, 1997, 362-363.
[118] Basha, AC; Soloman, PA; Velan, M; Miranda, LR; Balasubramanian, N; Siva, R. *J. Hazard. Mater.*, 2010, 176, 154-164.
[119] Daneshvar, N; Khataee, AR; Djafarzadeh, N. *J. Hazard. Mater.*, 2006, 137, 1788-1795.
[120] Daneshvar, N; Ashassi Sorkhabi, H; Kasiri, MB. *J. Hazard. Mater.*, 2004, 112, 55-62.
[121] Yousuf, M; Mollah, A; Schennach, R; Parga, JR; Cocke, DL. *J. Hazard. Mater.*, 2001, 84, 29-41.
[122] Chen, G. *Sep. Purif. Technol.*, 2004, 38, 11-41.
[123] Aber, S; Amani-Ghadim, AR; Mirzajani, V. *J. Hazard. Mater.*, 2009, 171, 484-490.
[124] Arul Murugan, A; Ramamurthy, T; Subramanian, B; Kannan, CS; Ganesan, M. *Int. J. Chem. Reactor Eng.*, 2009, 7, A83.
[125] Marselli, B; Garcia-Gomez, J; Michaud, PA; Rodrigo, MA; Comninellis, C. *J. Electrochem. Soc.*, 2003, 50, D79-D83.
[126] Drogui, P; Elmaleh, S; Rumeau, M; Bernard, C; Rambaud, A. *Water Res.*, 2001, 35, 3235-3241.
[127] Oturan, MA; Oturan, N; Lahitte, C; Trevin, S. *J. Electroanal. Chem.*, 2001, 507, 96-102.
[128] Gzmen, B; Oturan, MA; Oturan, N; Erbatur, O. *Environ. Sci. Technol.*, 2003, 37, 3716-3723.
[129] Hanna, K; Chiron, S; Oturan, MA. *Water Res.*, 2005, 39, 2763-2773.
[130] Irmak, S; Yavuz, SI; Erbatur, O. *Appl. Catal. B-Environ.*, 2006, 63, 243-248.
[131] Diagne, M; Oturan, N; Oturan, MA. *Chemosphere*, 2007, 66, 841-848.
[132] Brillas, E, Calpe, JC; Casado, J. *Water Res.*, 2000, 34, 2253-2262.
[133] Brillas, E; Boye, B; Sires, I; Garrido, JA; Rodriguez, RM; Arias, C; Cabot, PL; Comninellis, C. *Electrochim. Acta*, 2004, 49, 4487-4496.
[134] Brillas, E; Sauleda, R; Casado, J. *J. Electrochem. Soc.*, 1998, 145, 759-765.
[135] Martínez-Huitle, CA; Brillas, E. *Appl. Catal. B-Environ.*, 2009, 87, 105-145.

[136] Piuleac, CG; Rodrigo, MA; Canizares, P; Curteanu, S; Saez, C. *Environ. Model. Soft.*, 2010, 25, 74-81.
[137] Sadrzadeh, M; Mohammadi, T; Ivakpour, J; Kasiri, N. *Chem. Eng. J.*, 2008, 144, 431-441.
[138] Maier, HR; Morgan, N; Chow, CWK. *Environ. Model. Soft.*, 2004, 19, 485-494.
[139] Khataee, AR; Vatanpour, V; Amani, AR. *J. Hazard. Mater.*, 2009, 161, 1225-1233.
[140] Daneshvar, N; Salari, D; Khataee, AR. *J. Photochem. Photobio. A*, 2003, 157, 111-116.
[141] Daneshvar, N; Salari, D; Khataee, AR. *J. Photochem. Photobio. A*, 2004, 162, 317-322.
[142] Daneshvar, N; Aleboyeh, A; Khataee, AR. *Chemosphere*, 2005, 59, 761-767.
[143] Pons, MN; Alinsafi, A, Evenou, F; Abdulkarim, EM; Zahraa, O; Benhammou, A; Yaacoub, A; Nejmeddine, A. *Dyes Pigments*, 2007, 74, 439-445.
[144] Diebold, U. *Surf. Sci. Rep.*, 2003, 48, 53-229.
[145] Khataee, AR; Aleboyeh, H; A. Aleboyeh, A. *J. Experim. Nanosci.*, 2009, 4, 121-137.
[146] Khataee, AR; Pons, MN; Zahraa, O. *J. Hazard. Mater.*, 2009, 168, 451-457.
[147] Yates Jr, T; Thompson, TL. *Chem. Rev.*, 2006, 106, 4428-4453.
[148] Hoffmann, MR; Martin, ST; Choi, W; Bahnemannt, DW. *Chem. Rev.*, 1995, 95, 69-96.
[149] Daneshvar, N; Salari, D; Niaei, A; Khataee, AR. *J. Environ. Sci. Heal. B*, 2006, 41, 1273-1290.
[150] Blake, DM; Maness, PC; Huang, Z; Wolfrum, EJ; Huang, J. *Sep. Purif. Methods*, 1999, 28, 1-50.
[151] Tennakone, K; Wijayantha, KGU. *J. Photochem. Photobiol. A*, 1998, 113, 89-92.
[152] Calza, P; Sakkas, VA; Villioti, A; Massolino, C; Boti, V; Pelizzetti, E; Albanis, T. *Appl. Catal. B-Environ.*, 2008, 84, 379-388.
[153] Toma, FL; Guessasma, S; Klein, D; Montavon, G; Bertrand, G; Coddet, C. *J. Photochem. Photobio. A*, 2004, 165, 91-96.
[154] Duran, A; Monteagudo, JM. *Water Res.*, 2007, 4, 690-698.
[155] Monteagudo, JM; Duran, A; Guerra, J; Garcia-Pena, F; Coca, P. *Chemosphere*, 2008, 71, 161-167.

[156] Duran, A; Monteagudo, JM; Martin, IS; Garcia-Pena, F; Coca, P. *J. Hazard. Mater.*, 2007, 144, 132-139.
[157] Khataee, AR. *Environ. Technol.*, 2009, 30, 1155-1168.
[158] Oliveros, E; Benoit-Marquie, F; Puech-Costes, E; Maurette, MT; Nascimento, CAO. *Analusis*, 1998, 26, 326-332.
[159] Emilio, CA; Litter, MI; Magallanes, JF. *Helvetica Chimica Acta*, 2001, 84, 799-813
[160] Aleboyeh, A; Aleboyeh, H; Moussa, Y. *Dyes Pigments*, 2003, 57, 67-75.
[161] Gogate, PR; Pandit, AB. *Adv. Env. Res.*, 2004, 8, 553-597.
[162] Ibney Hai, F; Yamamoto, K; Fukushi, K. *Crit. Rev. Env. Sci. Technol.*, 2007, 37, 315-377.
[163] Brillas, E; Sires, I; Oturan, MA. *Chem. Rev.*, 2009, 19, 6570-6631.
[164] Degussa Corporation, Environmental Uses of Hydrogen peroxide (H_2O_2), Allendale, NJ, 2001.
[165] Venkatadri, R; Peters, RW. *Hazard. Waste Hazard. Mater.*, 1993, 10, 107-149.
[166] Rigg, T; Taylor, W; Weiss, J. *J. Chem. Phys.*, 1954, 22, 575-577.
[167] Huang, CP; Dong, C; Tang, Z. *Waste Manage.*, 1993, 13, 361-377.
[168] Kitis, M; Adams, CD; Daigger, GT. *Water Res.*, 1999, 33, 2561-2568.
[169] Yoon, J; Lee, Y; Kim, S. *Water Sci. Technol.*, 2001, 44, 15-21.
[170] Lu, MC; Lin, CJ; Liao, CH; Ting, WP; Huang, RY. *Water Sci. Technol.*, 2001, 44, 327-332.
[171] Buxton, GV; Greenstock, CL. *J. Phys. Chem. Ref. Data*, 1988, 17, 513-886.
[172] Walling, C; Goosen, A. *J. Am. Chem. Soc.*, 1973, 95, 2987-2991.
[173] De Laat, J; Gallard, H. *Environ. Sci. Technol.*, 1999, 33, 2726-2732.
[174] Bielski, BHJ; Cabelli, DE; Arudi, RL. *J. Phys. Chem. Ref. Data.*, 1985, 14, 1041-1100.
[175] Walling, C; Kato, S. *J. Am. Chem. Soc.*, 1971, 93, 4275-4281.
[176] Lin, SH; Lo, CC. *Water Res.*, 1997, 31, 2050-2056.
[177] Walling, C. *Acc. Chem. Res.*, 1975, 8, 125-131.
[178] Hickey, WJ; Arnold, SM; Harris, RF. *Environ. Sci. Technol.*, 1995, 29, 2083-2089.
[179] Lipczynska-Kochany, E; Sprah, G; Harms, S. *Chemosphere*, 1995, 30, 9-20.
[180] Tang, WZ; Tassos, S. *Water Res.*, 1997, 31, 1117-1125.
[181] Gob, S; Oliveros, E; Bossmann, SH; Braun, AM; Guardani, R; Nascimento, CAO. *Chem. Eng. Process.*, 1999, 38, 373-382.

[182] Khataee, AR; Vatanpour, V; Rastgar Farajzadeh, M. *Turkish J. Eng. Env. Sci.*, 2008, 32, 367-376.
[183] Stegemann, JA; Buenfeld, NR. *J. Hazard. Mater.*, 2002, 90, 169-188.
[184] Elmolla, ES; Chaudhuri, M; Eltoukhy, MM. *J. Hazard. Mater.*, 2010, 179, 127-134.
[185] Yu, RF; Chen, HW; Liu, KY; Cheng, WP; Hsieh, PH. *J. Chem. Technol. Biotechnol.*, 2010, 85, 267-278.
[186] Legrini, O; Oliveros, E; Braun, AM. *Chem. Rev.*, 1993, 93, 671-698.
[187] Costa, FAP; dos Reis, EM; Azevedo, JCR. J. Nozaki. *Sol. Energy.*, 2004, 77, 29-35.
[188] Collona, GM; Caronna, T; Marcandalli, B. *Dyes Pigments*, 1999, 41, 211-220.
[189] Ince, NH. *Water Res.*, 1999, 33, 1080-1084.
[190] Shen, YS; Wang, DK. *J. Hazard. Mater.*, 2002, 89, 267-277.
[191] Daneshvar, N; Khataee, AR. *J. Environ. Sci. Health.*, Part A, 2006, 41, 1-14.
[192] Durán, A; Monteagudo, JM; San Martin, I; Sánchez-Romero, R. *J. Hazard. Mater.*, 2009, 167, 885-891.
[193] Durán, A; Monteagudo, JM; Mohedano, M. *Appl. Catal. B-Environ.*, 2006, 65, 127-134.
[194] Monteagudo, JM; Durán, A; San Martin, I; Aguirre, M. *Appl. Catal. B-Environ.*, 2010, 95, 120-129.
[195] Giroto, JA; Guardani, R; Teixeira, ACSC; Nascimento,CAO. *Chem. Eng. Process.*, 2006, 45, 523-532.
[196] Nogueira, KRB; Teixeira, CSC; Nascimento, CAO; Guardani, R. *Braz. J. Chem. Eng.*, 2008, 25, 671-682.
[197] Kasiri, MB; Aleboyeh, H; Aleboyeh, A. *Appl. Catal. B-Environ.*, 2008, 84, 9-15.
[198] Feng, J; HU, X; Yue, PL. *Environ. Sci. Technol.*, 2004, 38, 269-275.
[199] Guzman-Vargas, A; Delahay, G; Coq, B; Lima, E; Bosch, P; Jumas, JC. *Catal. Today*, 2005, 107-108, 94-99.
[200] Yip, AC; Lam, FL; Hu, X. *Ind. Eng. Chem. Res.*, 2005, 44, 7983-7990.
[201] Kusic, H; Koprivanac, N; Selanec, L. *Chemosphere*, 2006, 65, 65-73.
[202] Kawai, T; Tsutsumi, K. *Colloid. Polym. Sci.*, 1995, 273, 787-792.
[203] Kusic, H; Bozic, AL; Koprivanac, M; Papic, S. *Dyes Pigments*, 2007, 74, 388-395.
[204] Durán, A; Monteagudo, JM; Amores, E. *Appl. Catal. B-Environ.*, 2008, 80, 42-50.

[205] Monteagudo, JM; Durán, A; Lopez-Almodovar, C. *Appl. Catal. B-Environ.*, 2008, 83, 46-55.
[206] Monteagudo, JM; Durán, A; San Martin, I; Aguirre, M. *Appl. Catal. B-Environ.*, 2009, 89, 510-518.
[207] Gogate, PR; Pandit, AB. *Adv. Environ. Res.*, 2004, 8, 553-597.
[208] Munter, R. Proc. Estonian Acad. Sci. Chem., 2001, 50, 59-80.
[209] Kasiri, MB; Aleboyeh, H; Aleboyeh, A. *Environ. Technol.*, 2010, 31, 165-173.
[210] Mohey, A; Libra, JA; Wiesmann, U. *Chemosphere*, 2003, 52, 1069-1077.
[211] Shu, HY; Chang, MC; Fan, HJ. *J. Hazard. Mater.*, 2004, 113, 201-208.
[212] Gogate, PR; Pandit, AB. *Adv. Environ. Res.*, 2004, 8, 501-551.
[213] Khataee, AR. *Pol. Chem. Technol.*, 2009, 11, 38-45.
[214] Khataee, AR. Kasiri, MB, *J. Mol. Catal. A: Chem.*, 2010, 331, 86-100.
[215] Salari, D; Daneshvar, N; Aghazadeh, F; Khataee, AR. *J. Hazard. Mater.*, 2005, 125, 205-210.
[216] Guimaraes, OLC; Silva, MB. *Chem. Eng. Process.*, 2007, 46, 45-51.
[217] Rice, RG; Robson, CM; Miller, GW; Hill, AGM. *J. Am. Water Works Assoc.*, 1981, 85, 63-72.
[218] Hoigne, J; Bader, H. *Water Res.*, 1976, 10, 377-386.
[219] Hoigne, J; Bader, H. *Water Res.*, 1983, 17, 185-194.
[220] Hoigne, J. *Water Sci. Technol.*, 1997, 35, 1-8.
[221] Prado, J; Arantegui, J; Chamarro, E; Esplugas, S. *Ozone-Sci. Eng.*, 1994, 16, 235-245.
[222] Poznyak, T; Chairez, I; Poznyak, A. *Water Res.*, 2005, 39, 2611-2620.
[223] Oguz, E; Tortum, A; Keskinler, B. *J. Hazard. Mater.*, 2008, 157, 455-463.
[224] Oguz, E; Keskinler, B; Tortum, A. *Chem. Eng. J.*, 2008, 141, 119-129.
[225] Legube, B; Parinet, B; Gelinet, K; Berne, F; Croue, JP. *Water Res.*, 2004, 38, 2185-2195.
[226] White, GC. *Handbook of chlorination and alternative disinfectants*, 4th ed. NY: John Wiley & Sons Inc., 1999, 212-85, 537-74.
[227] Wolfe, RL; Ward, NR; Olson, BH. *Environ. Sci. Technol.*, 1985, 19, 1192-1195.
[228] Kim, YH; Hensley, R. *Water Environ. Res.*, 1997, 69, 1008-1014.
[229] Devokta, LM; Williams, DS; Matta, JH; Albertson, OE. *Water Environ. Res.*, 2000, 72, 610-617.
[230] Yu, RF; Chen, HW; Cheng, WP; Shen, YC. *Resour. Conserv. Recy.*, 2008, 52, 1015-1021.

[231] Guimaraes, OLC; dos Reis Chagas, MH; Filho, DNV; Siqueira, AF; Filho, HJI; de Aquino, HOQ; Silva, MB. *Chem. Eng. J.,* 2008, 140, 71-76.
[232] Guimaraes, OLC; Filho, DNV; Siqueira, AF; Filho, HJI; Silva, MB. *Chem. Eng. J.,* 2008, 141, 35-41.

INDEX

A

abatement, 72
absorption, 31
abstraction, 67
accounting, 68
accuracy, 4, 27, 29
acid, 32, 60, 61, 62, 63, 65, 68, 76, 77
acquisition of knowledge, 27
activated carbon, 31, 32, 33, 34, 35
adaptation, 9, 10
adjustment, 30
adsorption, viii, 2, 31, 32, 33, 34, 35, 37, 43, 87
advantages, 28, 37, 38, 49
algae, 37
algorithm, 3, 9, 12, 14, 15, 16, 18, 20, 21, 25, 32, 34, 38, 39, 40, 43, 49, 54, 70, 71, 72, 74, 75, 76, 79, 80, 82, 83
alimentary canal, 32
alkenes, 68
amines, 80
ammonia, 38, 42, 45, 80
aniline, 54, 71, 74
aqueous solutions, 32, 43, 58, 71, 80
architecture, 17
aromatic compounds, 40, 43, 66
aromatic hydrocarbons, 31
aromatics, 41, 79
arsenic, 31
assessment, 40, 43
attachment, 62
authentication, 29
axons, 22

B

bacteria, 37, 58
band gap, 57
bankruptcy, 27
beneficial effect, 73
bias, 7, 17, 18
bicarbonate, 62
biodegradability, 43
biodegradation, 43, 44
biological processes, 38
biological sciences, 1
biological systems, 38
bonds, 68, 79
breast cancer, 30
bromine, 31
butyl ether, 82

C

cadmium, 31, 40
calculus, 92
cancer, 30, 32
carbon, 31, 34, 37, 51, 54, 72, 79
carbon materials, 31

catalyst, 62, 72
cheese, 44
chemical industry, 49
chemical stability, 57
chemometric techniques, 62, 63
China, 92, 94
chlorinated hydrocarbons, 58
chlorination, 65, 80, 81, 82, 84, 100
chlorine, 31, 53, 66, 80, 84
chromium, 31
class, 15
clinical trials, 30
clustering, 29
CO_2, 57, 58, 66, 68
coagulation process, 51, 54
coal, 32
coatings, 29
color, 22, 70, 75
combined effect, 87
combustion, 63
compatibility, 49
competition, 69, 78
competitive process, 11
compilation, 4
complexity, vii, 38, 59, 70, 81, 87
compounds, 2, 31, 32, 38, 43, 53, 54, 58, 66
compression, 30
computation, 29, 63
computer use, 29
computing, 16, 24
conduction, 57, 58
configuration, 43, 63, 70, 79
configurations, 49
conjugate gradient method, 15
consumer choice, 28
consumption, 78
contact time, 32
contaminant, 62
convergence, 12
cooling, 41
copper, 31, 43
correlation, vii, 19, 21, 33, 44, 70, 71, 72, 78, 79, 80, 87

correlation coefficient, 21, 33, 44, 70, 71, 72, 79, 80, 87
cost, 28, 31, 44, 50, 57
crystalline, 29, 30
cyanide, 64
cyanosis, 32
cycles, 63

D

data set, 9, 19, 24, 25, 28, 54, 71, 72, 78, 79, 80, 81
database, 29
decomposition, 66, 67, 68, 69, 80, 83
degradation, 41, 49, 51, 59, 60, 61, 62, 63, 70, 71, 72, 74
degradation process, 59, 63, 70
degradation rate, 71
Degussa, 62, 98
dendrites, 22
denitrification, 38
dependent variable, 3
deposition, 58
derivatives, 13, 15, 16
destruction, 37, 50, 57, 58, 65, 66
detection, 27
diagnosis, 30
diffusion, 51
disadvantages, 31, 72
discontinuity, 8
discrimination, 30
disinfection, 79, 80, 81
dissociation, 66, 68, 73
Doha, 40, 44
dosage, 81
drinking water, 32, 79, 80
dyes, 31, 54, 58, 70, 79
dynamic control, 81

E

effluent, 44, 45, 71, 75, 80
effluents, 49, 60, 72
electrodes, 52

electrolysis, 53, 54
electrolyte, 52, 53
electron, 49, 51, 57, 79
electrons, 57
energy consumption, 37, 49
engineering, 28, 30
equality, 13
equilibrium, 33, 35, 40
experimental condition, 38, 50, 64, 79, 87
experimental design, 62, 63
expert systems, 29
exploitation, 11
exploration, 11
extraction, 28, 30

F

fault detection, 30
feature detectors, 4
fermentation, 41
ferric ion, 67
ferrous ion, 68, 69
filters, 34
flexibility, 29, 31, 35, 38
flight, 30
fluid, 30, 40
fluidized bed, 43
fluorine, 31
forecasting, 28, 29
fraud, 27
free energy, 40, 43
free radicals, 68, 73
funding, 1
fungi, 37

G

gasification, 71
graph, 5, 19
graphite, 54

H

HE, 95

health problems, 32
heat transfer, 55
heavy metals, 58
hedging, 29
human brain, vii, 27
hybrid, 40, 79
hydrocarbons, 31, 58, 71
hydrogen, 51, 64, 66, 67, 68, 69, 71, 73, 78
hydrogen peroxide, 51, 64, 66, 67, 69, 71, 73, 78
hydroxide, 69
hydroxyl, 50, 57, 58, 66, 67, 68, 69, 70, 73, 78, 79
hysteresis, 30
hysteresis loop, 30

I

ideal, 21
image, 30
image analysis, 30
independent variable, 3
industrial wastes, 32
information retrieval, 29
initiation, 67
intervention, 11
intrinsic value, 4
iodine, 31
ions, 32, 39, 40, 43, 54, 67, 69, 70
Iran, viii, 89
iron, 66, 67
irradiation, 62, 63, 64, 70
iteration, 10, 12, 14, 15, 16, 24

J

job scheduling, 30

K

kinetic constants, 64
kinetic parameters, vii, 38, 59, 63
kinetics, 71, 73, 78

L

learning, vii, 3, 8, 9, 10, 11, 12, 41, 71, 79, 81, 82, 84, 91
learning process, vii, 3, 11
lens, 64
linear function, 7
linear model, 28
linear modeling, 28
litigation, 29

M

majority, 65
MANOVA, 62, 63
manufacturing, 2, 30
mapping, 3
market segment, 28
marketing, 27
matrix, 13, 14, 15, 16, 22, 30, 38
media, 62, 78
mercury, 31
metabolism, 37
metal salts, 66
methodology, 62
microorganism, 43
mixing, 84
modeling, vii, 1, 2, 20, 24, 29, 30, 32, 33, 34, 38, 39, 41, 43, 44, 45, 49, 50, 52, 54, 55, 59, 60, 62, 65, 70, 71, 72, 74, 78, 79, 80, 81, 82, 87
molds, 58
molecular oxygen, 58
molecular structure, 43
momentum, 43
monitoring, 70
multivariate calibration, 30

N

NaCl, 74, 82
nanocrystals, 29, 30
nanomaterials, 58
nanoparticles, 59, 62
nanotube, 54
neural network, vii, 1, 2, 3, 4, 5, 8, 9, 15, 17, 20, 21, 22, 23, 27, 28, 29, 32, 33, 38, 41, 43, 44, 45, 49, 50, 54, 55, 59, 62, 63, 64, 70, 71, 72, 78, 79, 80, 81, 87
Neural Network Model, i, iii, 27
neural networks, vii, 1, 2, 3, 15, 17, 28, 29, 32, 33, 38, 44, 45, 49, 55, 59, 62, 63, 64, 70, 78, 81, 87
neurons, vii, 1, 3, 4, 5, 9, 11, 17, 19, 20, 22, 32, 43, 44, 63, 64, 70, 79
nickel, 29, 30
nitrate, 32, 42, 45
nitrogen, 38, 63, 80
nodes, 3, 18, 19
noise, 30
nonlinear systems, 38
nutrients, 30
nutrition, 37

O

olefins, 79
optimization, 12, 15, 17, 27, 28, 30, 87
organic compounds, 31, 35, 38, 66, 70, 80
organic materials, 37
organic matter, 58, 73, 78
organizing, 3, 27, 53
oxidation, 2, 37, 50, 51, 52, 57, 65, 68, 69, 78, 79
oxygen, 42, 44, 45, 67, 70, 79
ozonation, 65, 79, 80, 82
ozone, 30, 66, 73, 79

P

parallel, 3
partition, 40, 43
pathways, 79
pattern recognition, vii, 1
performance, 8, 12, 15, 16, 18, 19, 20, 24, 30, 38, 44, 45, 49, 54, 62, 70, 71, 72, 78, 81
peroxide, 64, 66, 73, 98

pesticide, 34
pesticides, 31, 33, 34, 58
pH, 32, 34, 39, 40, 41, 43, 52, 53, 60, 61, 62, 63, 66, 67, 68, 69, 72, 74, 75, 76, 77, 78, 82, 83, 84
phenol, 39, 43, 53, 54, 72, 83
phosphorus, 41
photocatalysis, 57
photocatalysts, 57
photolysis, 65, 70, 73
photons, 57
physical properties, 31
pollution, 1, 30, 37, 50, 63
polymer, 30, 69
polymer composites, 30
polyvinyl alcohol, 72
population growth, 2
power plants, 63
probability, 78
process control, 70
prognosis, 30
programming, 91
propagation, 3, 8, 11, 12, 16, 18, 20, 21, 25, 32, 33, 34, 35, 38, 39, 40, 41, 42, 43, 49, 50, 52, 53, 54, 60, 61, 64, 68, 69, 70, 71, 72, 73, 74, 75, 76, 79, 80, 82, 83, 84
protons, 67
PTFE, 54
public health, 1
purity, 30
PVA, 76

R

radar, 30
radiation, 59, 70, 78
radical mechanism, 66
radicals, 57, 58, 65, 66, 67, 68, 70, 73, 80
reactant, 49
reactants, 68
reaction rate, 62, 80
reaction time, 62, 63
reactions, vii, 38, 49, 59, 67, 68, 69, 78, 79
real estate, 29
reasoning, 41

recognition, 27, 29, 30
recombination, 78
reconstruction, 30
red mud, 32
regeneration, 31
regression, 21, 32, 43, 50, 54, 72, 87
regression analysis, 21
reinforcement, 10, 11
reinforcement learning, 10
relevance, 64
remediation, 51, 58
residuals, 81, 84
resources, 2, 29, 87
rings, 68
river basins, 30
routines, 23

S

saccharin, 29
salinity, 30
salts, 66, 68, 78
sawdust, 39, 43
scaling, 20
scarcity, vii
scatter, 21
scatter plot, 21
scheduling, 30
selectivity, 80
sensitivity, 28, 43
sensors, 30
sequencing, 41
serum, 33, 35
serum albumin, 33, 35
signals, 4, 9, 19
simulation, viii, 2, 17, 18, 30, 44, 62, 81, 87
sludge, 40, 50
software, 29
solid surfaces, 37
species, 38, 39, 57, 58, 59, 65, 70, 78
speech, 1
standard deviation, 33, 80
statistics, 92
stoichiometry, 65

structural characteristics, 78
substrates, 68, 80
subtraction, 9
sulfites, 66
sulfur, 41
suppression, 30
surface area, 72
synapse, 22

T

temperature, 29, 32, 37, 41, 43, 44
testing, 44
topology, 2, 3, 17, 19, 20, 21, 25
total energy, 40, 43
toxicity, 38, 57
training, 2, 3, 12, 14, 15, 17, 18, 22, 23, 24, 25, 32, 33, 38, 43, 44, 50, 54, 63, 70, 71, 72, 78, 79
training speed, 15
transformation, 4
transition metal, 72
transport, 30
treatment methods, 57, 65, 87
trial, 10

U

ultrasound, 73
ultraviolet range, 57
uniform, 62
updating, 24
urbanization, vii
UV, 22, 57, 60, 61, 62, 65, 66, 70, 71, 73, 74, 75, 76, 78, 79, 82, 84
UV irradiation, 70
UV light, 57, 61, 62, 73, 82, 84
UV radiation, 62, 78

V

valence, 57
validation, 2, 24, 25, 38, 78, 79
variations, 11, 12
vector, 10, 12, 13, 16
vegetation, 37
velocity, 34
versatility, 2, 49
video, 30
viruses, 29, 58
viscosity, 41
vision, 1

W

waste, 32
wastewater, vii, 1, 2, 29, 30, 31, 32, 34, 37, 38, 39, 40, 41, 42, 43, 44, 45, 49, 50, 51, 52, 54, 57, 59, 60, 63, 65, 66, 70, 71, 74, 80, 81, 82, 84, 87, 95
water resources, vii, 2
workers, 27, 38, 50, 54, 63, 72, 79, 80
workplace, 29
worry, 23

Y

yeast, 41

Z

zeolites, 32, 72
zinc, 31, 40
ZnO, 57, 60, 63